UNDERSTANDING AND TROUBLESHOOTING DIGITAL ELECTRONIC CIRCUITS

UNDERSTANDING AND TROUBLESHOOTING DIGITAL ELECTRONIC CIRCUITS

JOHN DOUGLAS-YOUNG

PRENTICE HALL, Englewood Cliffs, New Jersey 07632

Library of Congress Cataloging-in-Publication Data

Douglas-Young, John.
   Understanding and troubleshooting digital electronic circuits /
John Douglas-Young.
     p. cm.
   Includes index.
   IBSN 0-13-932971-4
   1. Digital electronics.   I. Title.
TK7868.D5D68 1992

621.3815—dc20   91-31997
                                        CIP

Acquisition Editor: *George Kuredjian*
Production Editor: *Raeia Maes*
Copy Editor: *William Thomas*
Cover Designer: *Wanda Lubelska*
Prepress Buyer: *Mary McCartney*
Manufacturing Buyer: *Susan Brunke*
Supplements Editor: *Alice Dworkin*

 © 1992 by Prentice-Hall, Inc.
A Simon & Schuster Company
Englewood Cliffs, New Jersey 07632

Printed in the United States of America

10 9 8 7 6 5 4 3 2 1

ISBN   0-13-932971-4

PRENTICE-HALL INTERNATIONAL (UK) LIMITED, *London*
PRENTICE-HALL OF AUSTRALIA PTY. LIMITED, *Sydney*
PRENTICE-HALL CANADA INC., *Toronto*
PRENTICE-HALL HISPANOAMERICANA, S.A., *Mexico*
PRENTICE-HALL OF INDIA PRIVATE LIMITED, *New Delhi*
PRENTICE-HALL OF JAPAN, INC., *Tokyo*
SIMON & SCHUSTER ASIA PET. LTD., *Singapore*
EDITORA PRENTICE-HALL DO BRASIL, LTDA., *Rio de Janeiro*

Contents

PREFACE   *xi*

**1 BREADBOARDS AND THE COMPONENTS USED
WITH THEM**   **1**

Breadboards, 1

Fixed Resistors, 3

Variable Resistors, 6

Capacitors, 6

Diodes, 7

Transistors, 8

Integrated Circuits, 8

Handling Semiconductor Devices, 10

Other Precautions, 11

2 ELECTRONIC POWER SUPPLIES   **12**

Alternating Current, 12

Power Transformers, 13

Rectifiers, 14

Half-wave Rectification, 15

Filters, 16

Calculating the Value of the Filter Capacitor, 18

Surge Resistor, 18

Full-wave Rectification, 18

Full-wave Bridge Rectifier, 18

Calculating the Value of the Filter Capacitor, 20

Full-wave Center-tap Rectifier, 20

Regulation, 20

Zener Diodes, 21

Zener-Diode Shunt Regulator, 21

Series Transistor Regulator, 23

Series Regulator with Feedback, 24

Three-terminal Regulators, 25

Switching Regulator, 27

Troubleshooting Power Supplies, 28

3 THE FUNCTION GENERATOR   **33**

Function Generator IC, 34

Pulse Generator, 40

Modulation, 40

DC Output Level, 41

Troubleshooting Function Generators, 42

4 OPTOELECTRONIC DEVICES   **44**

Light-emitting Diodes, 44

LEDs in Digital Systems, 45

Binary Number Code, 45

Encoder, 47

Decoder, 49

Seven-segment Numeric Display, 49

Liquid-crystal Display, 51

Other Types of Display, 53

Displays with More Than One Character, 54

Optical Isolator, 55

Troubleshooting Optoelectronic Devices, 57

5 TIMING PULSE GENERATORS: CLOCKS   **60**

Quad Two-input NAND Gate (7400), 60

Timer 555, 63

Timer 555 Clock, 66

Op-amp Clock, 68

Using the 555 IC as a Voltage-controlled Oscillator, 68

Troubleshooting Clock Circuits, 69

6 DIGITAL LOGIC   **73**

Truth Tables, 75

Gates, 76

Digital Logic Families, 77

Transistor–Transistor Logic, 78

Emitter-coupled Logic (ECL), 79

Integrated Injection Logic, 79

Complementary Metal Oxide Semiconductor, 80

Logic Gate ICs, 82

Troubleshooting Tools, 82

Troubleshooting Digital Logic, 85

7 FLIP-FLOPS AND COUNTERS   **89**

R–S Latch, 89

Master–Slave Flip-flop, 91

D Flip-flop, 92

T Flip-flop, 92

J–K Flip-flop, 92

Asynchronous Binary Ripple Counter, 93

Synchronous Counter, 95

Decade Counters, 97

Integrated Counters, 98

Troubleshooting Counters, 99

8 MEMORY   **102**

Registers, 102

Larger Memories, 104

Multiplexer, 111

Demultiplexer, 113

Troubleshooting Registers, 114

9 ANALOG CIRCUITS   **115**

Differential Amplifier, 115

Operational Amplifier, 116

Inverting Voltage Amplifier, 118

Noninverting Voltage Amplifier, 119

Buffer, 120

Voltage Comparator, 120

Offset Compensation, 121

Frequency Compensation, 121

Difference Amplifier, 122

Summing Amplifier, 122

Logarithmic Amplifier, 123

Antilog Amplifier, 124

Voltage Multiplier and Divider, 124

Exponientation and Extraction of Roots, 125

Wave Generation and Shaping, 125

Square-wave Generator, 125

Sinusoidal Oscillator, 126

Pulse Generator, 126

Function Generators, 128

Active Filters, 128

Low-pass Filter, 128

High-pass Filter, 128

Bandpass Filter, 128

Band-reject or Notch Filter, 131

Troubleshooting Op-amps, 133

**10 ANALOG–DIGITAL AND DIGITAL–ANALOG
CONVERSION**   **134**

A/D Conversion, 134

D/A Conversion, 140

Resistive-ladder Network, 141

Troubleshooting A/D and D/A Convertors, 146

APPENDIX A: SEMICONDUCTORS   **147**

APPENDIX B: DIODES AND TRANSISTORS   **149**

**APPENDIX C: CONVERTING A BREADBOARD
EXPERIMENT TO A PERMANENT FORM**   **153**

**GLOSSARY OF ELECTRONIC TERMS USED IN
THIS BOOK**   **155**

INDEX   **165**

Preface

This book is about digital electronics. It covers a broad spectrum of basic circuits (and supporting analog circuits) to help you, the digital electronics technician, to troubleshoot with simple, but effective, methods, using low-cost test equipment, such as logic probes, pulsers, digital multimeters, and more.

In this book we cover a wide range of basic and practical applications of digital circuits, together with power supplies and operational amplifiers. Most of these circuits are *integrated circuits* (ICs) or "microchips." Their internal circuitry is not visible, so their description, for the most part, is in the form of block diagrams.

In Chapter 1 we show how to use the solderless circuit board (breadboard), which can be very useful in troubleshooting, and we review components that are used with ICs: resistors, capacitors, diodes, transistors, and so on.

In Chapter 2 we examine the circuits used in power supplies to change ac from the power line to the dc needed to operate electronic equipment. These range from the simplest half-wave rectifier to the most modern switching regulator.

In Chapter 3 we discuss a function generator, using a popular integrated circuit. We also show how you can build this easily into a permanent piece of test equipment if you need a source of square, sine, and triangle waves for future troubleshooting.

In Chapter 4 we see how various optoelectronic devices work. These provide the visual means through which digital equipment "talks" to us and tells us what is going on. We look at light-emitting diodes, liquid-crystal displays, and the integrated circuits that control them.

Proper operation of all but the least sophisticated digital equipment requires that the digital circuits be precisely synchronized with each other, so in Chapter 5 we go on to investigate "clocks," which are the circuits that generate timing pulses for this purpose.

In Chapter 6 we review digital logic, which is the basis of digital electronics, and the logic gates that we use to implement it. We also discuss troubleshooting procedures for integrated circuits and the best test equipment to use.

In Chapter 7 we investigate flip-flops, latches of all kinds, and counters. These are the building blocks of many digital devices used in computers, calculators, and modern audio equipment.

Chapter 8 is devoted to memory devices. This chapter covers registers, memories, and tristate buffers and how a microprocessor writes data into memory and reads it out. Multiplexers and demultiplexers are also included.

The operational amplifier is the workhorse of modern electronics. Although it is an analog device, it is an essential part of many digital circuits, so we cover it in Chapter 9. Although the op-amp is an amplifier, we shall see that it may also be used as an oscillator, comparator, buffer, adder, subtracter, multiplier, divider, and active filter.

Chapter 10 covers analog-to-digital (A/D) and digital-to-analog (D/A) devices, which are used in modern digital audio systems, and shows how they are linked together to process an audio signal.

A certain amount of math is used in explaining how these circuits work. This is given to *prove* the statements made and the results obtained, but you do not have to commit it to memory. You will probably never use it in practice.

Questions and answers are scattered throughout the text. These are intended to draw your attention to various aspects of the subject matter, not to duplicate it, but to strengthen your understanding of the subject being discussed.

There are also many illustrations in this book, on the principle that "a picture is worth a thousand words."

Lastly, there are three appendices. Appendix A reviews briefly how semiconductors work. Appendix B shows how semiconductors are used in diodes and transistors and shows the way bipolar junction transistors differ from field-effect transistors. Appendix C shows how you can convert your function generator breadboard experiment into a permanent test instrument for your work bench.

The appendices are followed by a Glossary of the electronic terms used in this book, so that you won't have to hunt for expressions or abbreviations used earlier but that momentarily do not come to mind.

Troubleshooting digital equipment can be, at the very least, frustrating, if not almost impossible, when done by hit or miss methods. To diagnose efficiently the trouble with a "sick" digital circuit, you must know how it behaves when it is "well." The circuits in this book are shown in good health so that when you encounter one with a problem you will be able to put your finger with certainty on the cause by comparing its current behavior with what it should be. You can also build similar circuits on your breadboard and deliberately introduce defects to see how they affect circuit performance.

You must know already that efficiency in troubleshooting is one of the most valuable accomplishments you can have. In fact, you are the most important person around when things go wrong. (And, as Murphy's law says, if they can, they will.) This book was written to help you become that person.

In conclusion, I wish to acknowledge my indebtedness to Charles L. Howard, Matsushita Service Co., and Robert Mendelson, RCA (retired), who generously gave of their time to read the original manuscript and who made many valuable suggestions for its improvement.

John Douglas-Young

UNDERSTANDING AND TROUBLESHOOTING DIGITAL ELECTRONIC CIRCUITS

1

Breadboards and the Components Used with Them

BREADBOARDS

What is a breadboard? This is a purely rhetorical question; you don't have to answer it! I know you know it's got nothing to do with bread or dough, although in the early days we used similar wooden boards for equipment mockups; hence the name.

No, the breadboard we are talking about is a *solderless breadboard*, meaning a device on which we can lay out electronic circuits to test them without having to solder any wires or components. This is a great idea, since many components are susceptible to damage by heat. Furthermore, we can rearrange the connections without having to desolder them. In fact, it is as simple as child's play, and the parts can be used over and over again, just like Lego.

Figure 1-1(a) shows the upper side of a simple breadboard that you can get from Radio Shack, and Figure 1-1(b) shows its bottom side with the base removed. There are 856 holes in the upper plastic surface of the board, and they are joined together underneath by 128 short metal strips and 8 long ones. Actually, the metal strips are U-shaped channels. The figure shows their underneaths only.

The sides of these channels, which we cannot see, are springy, so that if we push a #22 AWG solid wire down through one of the holes in the top of the board it will force the channel's sides apart and then be held in position by them until we pull it out again.

Question: We said a #22 AWG solid wire. Why? What would happen if we used a thinner wire (say #30 AWG) or a thicker wire (say #18 AWG)? [Well, the thinner wire would not make a reliable contact with

1

Figure 1-1(a)   Simple breadboard (upper side) with a full-wave regulated 5-V power supply.

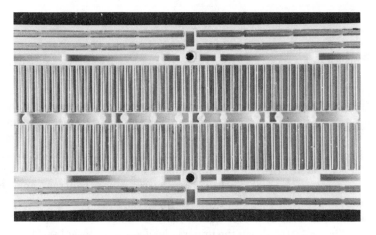

Figure 1-1(b)   Simple breadboard (underneath).

the metal channel, and a thicker wire would spread the sides of the channel too much, and might permanently damage it, don't you think?]

We also specified solid wire. Stranded wire is too flexible and would be difficult to force down into the holes. We need a supply of #22 AWG insulated wire in assorted colors, and it's a good idea to prepare a number of pieces of different lengths, with the insulation removed from about a quarter of an inch at each end.*

Anyway, the channels connect together rows of holes. If two wires are inserted into holes connected by a common channel, they are effectively joined together

---

* Prepared wires are available from some electronic suppliers.

without the need for soldering. The shorter channels are used to connect components; the longer ones are for buses carrying common supply voltages or grounds.

A word of caution: The top surface of the breadboard is made of a plastic that will melt if exposed to heat. We must bear this in mind and avoid overheating anything in contact with it. Of course, we are not going to do any soldering in its vicinity, but a short circuit could cause a component to get extremely hot.

The top surface of the board has a groove down the middle. This is where we mount an integrated circuit (IC) of the dual-in-line package (DIP) type. The pins on one side of the DIP are inserted in the holes on one side of the groove, and those on the other side in the holes on the other side of the groove. The holes are spaced $\frac{1}{10}$ inch apart, which is standard for DIPs. Each DIP pin goes into the end hole of a short channel with five holes, so each pin can be connected directly to up to four other circuit elements. We can, of course, connect it to more by running a lead from one of these holes to another short channel elsewhere on the board.

> *Question:* Could we install a chip crosswise instead of lengthwise on the board? [No, because we would just be connecting the pins together. Also, unless the DIP was quite small, there would be no holes for those pins that coincided with the groove.]

Incidentally, mounting DIPs on the breadboard needs a little care. The pins are easily bent. We must get them all lined up correctly with their respective holes, and then push gently on the middle of the DIP to press them all in. When this has been done properly, the whole underneath of the DIP will be in contact with the surface of the board.

Removing a DIP also requires care. An IC extractor is the best tool for this, but we can also do it by inserting a small screwdriver under one end of the chip and levering it up slightly and then doing the same at the other end, repeating the process alternately at each end until it is free.

We shall have a few more things to say about DIPs later in this chapter. In the following section we shall explain component identification.

FIXED RESISTORS

Most resistors are made of a material (like carbon) that has a specified resistance to the flow of electric current. The value of such a resistor is given by colored bands around its body, as shown in Figure 1-2.

When there are four of these bands, one of them will be gold or silver. The other three will be various colors. We read them from the end opposite the gold or silver band. If there is no gold or silver band, there are only three bands, and they will be grouped closer to one end than the other. In this case, we start with the band nearest that end.

The colors of the bands have numerical meanings, as shown in Table 1-1. The

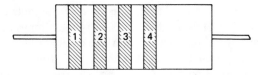

Figure 1-2 How value is marked on a resistor.

first two bands give the first two figures of the value; the third band gives the number of zeros following them. Thus, if the first band is brown (1), the second black (0), and the third brown (1), the resistor value is 1, 0, followed by one zero = 100. If the fourth band is gold, this value is accurate within ±5%.

The sizes of these resistors depend on their wattage ratings. Normal wattages are $\frac{1}{8}, \frac{1}{4}, \frac{1}{2}$, 1, and 2 watts, with voltage ratings of 150, 250, 350, and 500 volts. The most plentiful and cheapest are $\frac{1}{4}$-watt resistors, which are suitable for the majority of circuits.

> *Question:* Why do resistors have such oddball values? [Certain preferred values are used for resistors. Because of their allowed tolerances, it is not necessary to provide them in all values. For example, 20% resistors are made with values of 10, 15, 22, 33, 47, and 68 for those under 100 ohms; 100, 150, 220, 330, 470, and 680 for those from 100 to 999 ohms; and so on. Table 1-2 gives the two significant figures for the values for the three tolerances generally used.]

Occasionally, resistors show up with more than four colored bands. These are not ordinary carbon composition resistors of the type described so far, but are

TABLE 1-1   RESISTOR COLOR CODE

Color	Value
Black	0
Brown	1
Red	2
Orange	3
Yellow	4
Green	5
Blue	6
Violet	7
Gray	8
White	9
Gold	±5%
Silver	±10%
No color	±20%

TABLE 1-2 SIGNIFICANT FIGURES FOR
PREFERRED VALUES OF RESISTORS

5%	10%	20%
10	10	10
11		
12	12	
13		
15	15	15
16		
18	18	
20		
22	22	22
24		
27	27	
30		
33	33	33
36		
39	39	
43		
47	47	47
51		
56	56	
62		
68	68	68
75		
82	82	
91		
100	100	100

made to more exacting tolerances for use in test equipment or for military purposes. The first three bands may be significant figures, or the extra fifth band may indicate some special characteristic. Other colors may be used instead of gold or silver for the tolerances given in Table 1-2.

There are also larger resistors for power-handling purposes. These are made of wire wound on a ceramic tube, and their values are stamped on the body instead of using a color code. Instead of axial leads, these resistors have terminal lugs at each end and often have a third, sliding terminal for selecting a tap.

VARIABLE RESISTORS

Variable resistors are made for use as user controls or for internal adjustments of resistance. There are two types: those with two terminals and those with three. The first is called a rheostat, the second a potentiometer ("pot"). Their values are usually stamped on them. They may be composition, metallic film, or wirewound, the latter being used for higher currents.

While some miniature variable resistors have pins that fit the holes in the breadboard, many do not. If necessary, solder short lengths of #22 AWG hookup wire to their terminals, and insert these leads into the board. It might be desirable to support larger controls by mounting brackets.

CAPACITORS

Larger capacitors, such as electrolytic capacitors, have their values printed on them in microfarads. They often have arrows indicating the negative terminal or a groove for the positive terminal (or both). They will also state their maximum working voltage.

> ***Question:***   Why do we want to know which are the positive and negative terminals of an electrolytic capacitor? [In this type, the dielectric is a film of oxide electrolytically deposited on an electrode of aluminum or tantalum, which gives a high capacitance, but acts as a dielectric in one direction only. The device is therefore polarized and may be damaged if connected up with the wrong polarity. A nonpolarized electrolytic capacitor is really two such capacitors in series with their like terminals connected together.]

Some smaller tantalum electrolytics use a color code, as shown in Figure 1-3. The capacitance is then given in picofarads, in accordance with Table 1-3.

Small ceramic capacitors have their values stamped on them in typographical form, but they are expressed in picofarads in the style shown in Figure 1-4. The first two numbers are the significant figures in the value; the third is the number of zeros to add. For instance, if the number stamped on the capacitor is 100, it would mean 10 picofarads. But if it is 103, it means 10,000 picofarads, or 0.01 microfarad.

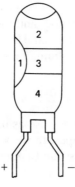

Figure 1-3   Small tantalum capacitor (see Table 1-3) (area 1 also indicates positive lead; area at top may show tolerance).

TABLE 1-3 COLOR CODE FOR SMALL TANTALUM CAPACITORS

Color	Voltage (1)	Value (2, 3)	Multiplier (4)
Black	4	0	—
Brown	6	1	—
Red	10	2	—
Orange	15	3	—
Yellow	20	4	10,000
Green	25	5	100,000
Blue	35	6	1,000,000
Violet	50	7	10,000,000
Gray	—	8	—
White	—	9	—

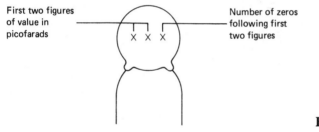

First two figures of value in picofarads

Number of zeros following first two figures

Figure 1-4 Small ceramic capacitor.

DIODES

A general-purpose diode has an indication on its body that denotes which of its two leads is the cathode. This may consist of a tapered end or a colored band of some kind. Germanium diodes, with bodies made of glass, use color coding to show their type. The colors have the same meaning as the colors used for marking resistors. Such diodes have identification numbers in the form 1N1234. The 1N is ignored, since it is the same on all. Four color bands give the rest of the number (in this case brown, red, orange, and yellow). Sometimes a fifth band is present. This represents a letter subscript, using brown through white for A through J, instead of the numbers 1 through 9.

Zener diodes may also have similar color coding. They are explained in Chapter 2.

The cathode of a light-emitting diode (LED) is either the shorter of its leads or the one next to a flat on its base. These are discussed in more detail in Chapter 3.

TRANSISTORS

Transistor packaging arrangements are numerous. The most widely used are illustrated in Figure 1-5.

> *Question:* Is there any difference between the packaging of a bipolar and a field-effect transistor? [Generally, no. The most common types use the TO92 plastic case, and they look exactly alike. This package is also used for unijunction transistors and silicon-controlled rectifiers.]

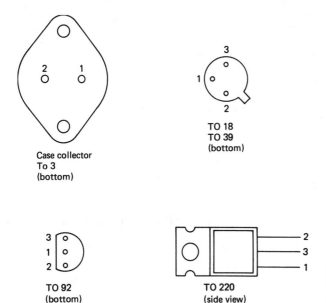

Case collector
To 3
(bottom)

TO 18
TO 39
(bottom)

TO 92
(bottom)

TO 220
(side view)

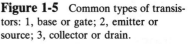

Figure 1-5 Common types of transistors: 1, base or gate; 2, emitter or source; 3, collector or drain.

INTEGRATED CIRCUITS

All integrated circuits (ICs) have their identification codes printed on them. There is a recommended, or preferred, referencing code, which is shown in Figure 1-6. However, manufacturers use it in different ways, insert other letters of their own, and often add their own ID number as well. ICs in metal cans that look like transistors with more than the usual number of leads will have very different numbers, even though the device is the same.

Figure 1-7 shows a typical dual-in-line package (DIP), which is the most popular type of package, with a typical set of identification numbers.

In addition to DIPs and cans, ICs are also supplied in flatpacks and flipchips. Flatpacks are also encapsulated in much the same way as DIPs, but their leads stick straight out from the sides. Flipchips have solder bumps instead of leads and are

```
                        XXAANNNNBCD

where:

  XX = manufacturer's identification (e.g., MC for MOTOROLA)

  AA = device family

          AD = analog-to-digital
          AH = analog hybrid
          AM = analog monolithic
          CD = CMOS digital
          DA = digital-to-analog
          DM = digital monolithic
          LF = linear FET
          LH = linear hybrid
          LM = linear monolithic
          LX = transducer
          MM = MOS monolithic

 NNN = device number

  B = device number suffix (if any)

  B = package style

          D = glass/metal DIP
          F = glass/metal flatpack
          H = small metal can
          I = glass/ceramic DIP
          K = large metal can
          L = small multipin metal can
          N = plastic DIP
          P = plastic can with heat sink
          S = power DIP
          T = metal can with heat sink
          W = flatpack
          Z = small plastic can

  D = temperature data
```

Figure 1-6 Preferred reference code for ICs.

mounted face down. We shall not consider these others here, since they cannot be plugged into the breadboard.

All ICs need some means of identifying their numerous leads. DIPs have a pin locator in the form of a dot or notch placed adjacent to pin #1 or at the same end of the case; or it may be in the form of a groove on the same side as pin #1. The pins are numbered counterclockwise from pin #1 as seen from above. Don't be fooled by DIPs that have a dot at one end and a notch at the other; only the notch counts!

Another way of identifying pin #1 is to view the DIP from above, so that the writing on it is the right way up. Pin #1 is then at the lower left.

An IC in a can has a tab sticking out over the pin with the highest number, so

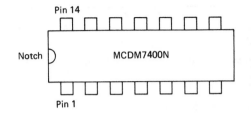

Figure 1-7 This is an imaginary IC. The reference tells us that it was manufactured by Motorola, is a digital monolithic device, number 7400, and is packaged as a plastic DIP.

that, as viewed from above, pin #1 is to the left of the tab, and the others are numbered counterclockwise around the can.

HANDLING SEMICONDUCTOR DEVICES

Most modern semiconductor devices are rugged enough that they can withstand reasonable mechanical shocks. However, they can be damaged by excessively rough use, especially those in metal cans, such as dropping on the floor. This also includes the force exerted by the tool used to cut the leads. If this tool causes the cut portion to fly off with a sharp snap, it transmits the same shock back along the lead to the device it comes from, which may be enough to break it off inside the case. It is best to use a scissor-type cutting tool for this, rather than a crimping type.

Soldering is an operation fraught with danger. We don't encounter this problem when using a breadboard, of course, but we are sure to encounter it when working on a printed circuit board. The following precautions should always be observed:

1. Solder as far as possible from the body of the device.
2. Don't apply heat or molten solder to a lead or terminal for more than 10 seconds at the most, or to a point nearer than 2 millimeters from the body of the device.
3. Use a low-voltage soldering iron, one that is 30 watts or less.
4. Make sure the surfaces to be soldered are clean and the tip of the iron tinned so that the operation can be done as quickly as possible.
5. Always grip the lead or terminal with long-nose pliers at a point between the body of the device and the soldering point. The pliers act as a heat sink, absorbing much of the heat before it reaches the device. If pliers are not available or you don't have a hand free to hold them, use a hemostat* or an alligator clip.

---

* A hemostat, a clamp used by surgeons for stopping the flow of blood, is a very useful tool; it is sold in electronic parts stores under the name "locking forceps."

OTHER PRECAUTIONS

Always turn off the power before inserting or unplugging any lead or device, since sudden transients may cause damage. This rule should be applied to all types of semiconductor devices, but especially to MOS devices. The latter are very susceptible to static charges. Very high static voltages can build up on the human body. We can protect semiconductor devices by doing the following:

1. Never remove or insert an MOS device with the power on.
2. Never apply signals to the device input with the power off.
3. Connect all unused leads to V_{SS} or V_{DD}, as appropriate.
4. Ground soldering-iron tips, metal tools, and so on, and wear a metal bracelet* that is also grounded.
5. When not actually mounted in a circuit, store MOS devices with their pins or leads in contact with a conductive material. Conductive foam is very good for this purpose.

Note: Most soldering irons are ac. However, you should use a dc iron for soldering MOS devices for the greatest safety.

---

* For safety, a resistor should be in the ground wire to avoid shock hazard from accidental contact with the ac line.

2

Electronic

Power Supplies

ALTERNATING CURRENT

Although many circuits with low current requirements do not need anything more than a battery (ignoring more sophisticated sources such as solar cells), a great range of electronic equipment has electronic power supplies that convert ac to dc to take advantage of the ac power supplied by the electric utility.

This power is generated mostly by alternators connected directly to steam turbines. The output of these machines is current and voltage alternating as shown in Figure 2-1. In this diagram, time is measured horizontally in degrees, corresponding to the angle of rotation of the armature of an alternator revolving at 3600 rpm, or 60 revolutions per second. (This is why the ac frequency is 60 hertz.) Current or voltage at each instant is measured vertically and varies in accordance with the sine of the angle of rotation; hence the term sine wave. However, most of these waves are not exactly sine waves, so they are more often termed *sinusoidal waves*. The practical difference is negligible.

The ac line voltage is nominally 120 V_{rms}, although it varies during the day according to the load on the system. The subscript *rms* indicates that this voltage is its *root-mean-square value*, meaning the value of the square root of the average of the squares of all the instantaneous values in one complete cycle of the sine wave. A potential of 1 V_{rms} is equal to the peak value $V_m \times 0.707$ and has the same power as 1 V_{dc}. From this you can see that a 120-V_{rms} sine wave will have a peak value of $120/0.707 = 170$ V_{pk}. The same relationship applies to current values.

> *Question:* Does the conversion factor 0.707 apply only to sine waves?
> [Yes. Ordinary ac meters generally give the correct rms values for sine waves, but not for other waveforms. For these you need a *true rms* meter.]

12

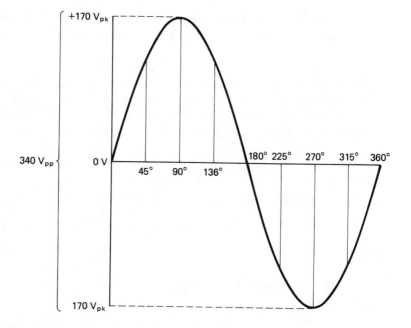

Figure 2-1   Sine wave of the electric utility. Its voltage is usually expressed as 120 V_{rms} (see text).

POWER TRANSFORMERS

One of the most important advantages of the use of ac is that its voltage can easily be changed by a transformer, something that cannot be done with dc. A power transformer also provides isolation from the ac line, an important consideration to minimize the risk of electric shock.

> *Question:*   Why won't a transformer change dc voltage? [A current is induced in the secondary winding only when the magnetic lines of force created by the current in the primary winding are *changing*. Those produced by steady dc are not changing, so no current is induced in the secondary.]

Voltages of 5 or 12 V_{dc}, and so on, are commonly used by electronic equipment, and transformers are available to lower the 120 V_{rms} of the power line to 6.3, 12.6 V_{rms}, or whatever level is wanted. In some cases, of course, it may have to be raised.

A power transformer has at least two windings, a primary and one or more secondaries, wound on the same iron core. The primary is connected across the power line, and the changed voltage appears across the secondary.

In choosing a power transformer, it is necessary to know how much current will be drawn from the secondary and to select a transformer whose secondary can handle the maximum load.

Warning: In the following exercise, 120 V$_{rms}$, which can give you a severe shock, will be present on the primary connections of the transformer.

To demonstrate this on your breadboard, install a small transformer, such as the one shown in Figure 2-2, so that its primary is connected to the power line. Then connect various resistors across the secondary and measure the ac current. Resistors with larger values allow less current to pass, but those with lower values result in higher currents. Some readings are shown in Table 2-1 (V$_{sec}$ = 8 V$_{rms}$).

Question: When we have changed the line voltage to the voltage we want to use in our circuit, is that it? [No. AC must be changed to dc before it can be used by most electronic equipment.]

RECTIFIERS

The current in the secondary of the transformer is ac. Two further steps are necessary to convert it to dc:

1. Make it flow in one direction only.
2. Smooth it, so that its voltage is level.

Making it flow in one direction only can be done in either of two ways. You can block all the negative peaks of the sine wave, leaving only the positive peaks, which is called *half-wave rectification*. Or you can reverse the polarity of the negative peaks so that all the peaks are positive. This is called *full-wave rectification* and is more efficient.

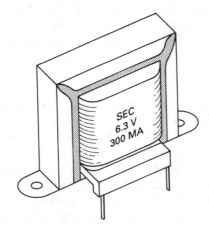

Figure 2-2 Small power transformer.

TABLE 2-1   TRANSFORMER SECONDARY CURRENTS WITH VARIOUS LOADS

Resistor Value (kΩ)	Current (mA)	Resistor Value (Ω)	Current (mA)
10.0	0.8	1000	8.0
2.2	3.6	220	36.3

HALF-WAVE RECTIFICATION

To block the negative peaks, a single silicon diode, as shown in Figure 2-3, may be used, with its anode connected to one end of the transformer secondary winding. The other end of the diode, with a light band or other distinguishing feature, is the cathode. Connected in this way, the diode blocks the negative peaks, but passes the positive ones.

The important diode parameters are the following:

1. Maximum peak inverse voltage, PIV
2. Maximum forward voltage drop, V_F
3. Maximum forward current, I_F
4. Maximum surge current
5. Maximum reverse current at PIV

PIV is the greatest peak voltage that the diode can withstand in the reverse direction. In Figure 2-4 we have used a 1N4003 diode, for which the maximum allowable value for this parameter is 200 V_{pk}. The maximum allowable value of V_F is 1.6 V. The maximum allowable value of I_F is 1 A. Maximum allowable surge current is 30 A for 16 ms (milliseconds). The maximum reverse current at PIV is 10 μA (microamperes).

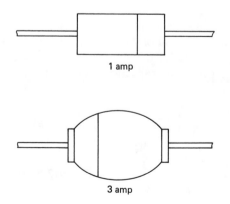

1 amp

3 amp

Figure 2-3   Typical silicon diodes used as half-wave rectifiers.

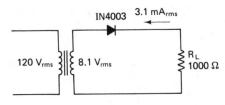

Figure 2-4 Half-wave rectifier.

The cathode of the diode is connected to a 1000-Ω resistor, which is the load resistor R_L, substituting for the real load. When the circuit is connected to the 120-V_{rms} line, 8.0 V_{rms} appears across the transformer's secondary.

In the circuit diagram in Figure 2-4, the diode blocks the negative halves of the sine wave appearing across the transformer secondary. In our exercise, the secondary voltage was measured on a digital multimeter (DMM) set to read ac volts. It was 8.1 V_{rms}; so with the negative halves of the sine wave removed it would be 4.05 V_{rms}. The voltage dropped across R_L was measured as 3.7 V_{rms}, so the voltage dropped across the diode was 4.05 − 3.7 = 0.35 V_{rms}. Since the actual (measured) resistance of R_L was 988 Ω, the circuit current had to be 3.7/988 = 3.7 mA_{rms}.

Question: When the DMM was switched to read dc volts, the voltage measured across R_L was 3.3 V_{dc}, giving a circuit current of 3.3 mA_{dc}. What does this mean? [It means that the DMM dc volts function sees the positive halves of the rectified sine wave as pulsating dc and indicates their *average* value, which is V_{rms} × 0.9, or 3.3 V. These pulsations are called *ripple*, which is undesirable.]

FILTERS

To reduce the ripple, the pulsating dc is filtered. In Figure 2-5 the *filter capacitor* has a value C of 100 μF (microfarads). The positive halves of the sine wave coming from the diode charge it to their peak value. The peak value of 8.1 V_{rms} is 8.1/0.707 = 11.5 V_{pk}.

For a 60-Hz (hertz) sine wave such as this, the total period of each cycle is 16.7 ms. When V_{sec} is larger than the capacitor voltage V_c, the diode conducts and charges the capacitor to 11.5 V_{dc}. Typically, the diode conducts for 3.7 ms. During the remaining 13 ms (t) of the cycle, the capacitor provides the load current by itself,

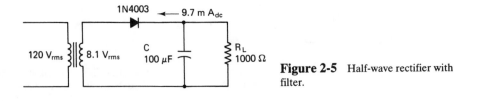

Figure 2-5 Half-wave rectifier with filter.

and V_c decreases linearly as shown in Figure 2-6. Our DMM measured the average V_c as 9.6 V_{dc}.

Question: Which circuit (Figure 2-4 or 2-5) had the highest dc output and the lowest ac output? [Figure 2-5.]

The circuit current I_L is given by V_{dc}/R_L = 9.6/988 = 9.7 mA$_{dc}$. t = 13 ms. The ripple voltage is given by

$$V_r = \frac{I_L \times t}{C}$$

$$= \frac{9.7 \times 10^{-3} \times 13 \times 10^{-3}}{100 \times 10^{-6}} \tag{2-1}$$

$$= 1.26 \, V_{pp}$$

$$= 0.4 \, V_{rms}$$

The DMM confirmed this by measuring 0.4 V_{rms} across the filter capacitor.
    The *ripple factor* r is given by V_r/V_{dc}. The ripple factor is also given by

$$r = \frac{\sqrt{2}}{2\pi f_r R_L C} \tag{2-2}$$

where f_r is the ripple frequency in hertz. This is 60 Hz in a half-wave rectifier and 120 Hz in a full-wave rectifier. Substituting in formula (2-2),

$$r = \frac{\sqrt{2}}{6.28 \times 60 \times 988 \times 100 \times 10^{-6}}$$

$$= 0.04$$

Since $r = V_r/V_{dc}$ = 0.4/9.6 = 0.04, you can see that we have another confirmation of the value of V_r.

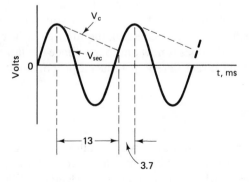

Figure 2-6 V_c represents charging and discharging of C in Figure 2-5.

CALCULATING THE VALUE OF THE FILTER CAPACITOR

The formula $r = \sqrt{2}/(2\pi f_r R_L C)$ may be rearranged to calculate the value for the filter capacitor. In this case, assume you want a ripple factor of 0.04:

$$C = \frac{\sqrt{2}}{2\pi f_r R_L r}$$

$$= \frac{\sqrt{2}}{6.28 \times 60 \times 988 \times 0.04} \qquad (2\text{-}3)$$

$$= 95 \ \mu\text{F} \cong 100 \ \mu\text{F}$$

SURGE RESISTOR

In selecting a diode, it is necessary to consider the surge current. This is the current through the diode when V_{sec} is first applied. The uncharged capacitor is like a dead short, so for a moment the diode current will be very high. The larger the capacitance is the longer this condition will last. In some cases it may be necessary to connect a low-value resistor between the diode and the capacitor to limit the surge current if the diode is not to be damaged.

In this circuit, there is not much likelihood of damage, because a 1N4003 diode can withstand a surge current of 30 A. However, if we were to insert a 5-Ω resistor between the diode and the capacitor, it would limit the secondary current I_{sec} as follows (ignoring the voltage drop across the diode):

$$I_{\text{sec}} = \frac{V_{\text{pk}}}{R} = \frac{11.5}{5} = 2.3 \ \text{A} \qquad (2\text{-}4)$$

FULL-WAVE RECTIFICATION

As already mentioned, a full-wave rectifier reverses the polarity of the alternate half of a sine-wave cycle. The sine wave consequently becomes pulsating dc with 120 peaks per second, instead of the 60 peaks per second you get from a half-wave rectifier. This reduces each period for recharging the filter capacitor to 3.3 ms, but the period during which it discharges is now only 5 ms. Since the latter period is less than half that of the half-wave rectifier, V_r is reduced proportionately.

FULL-WAVE BRIDGE RECTIFIER

Figure 1-1(a) showed a breadboard assembly of a full-wave bridge rectifier power supply, and Figure 2-7 gives the circuit diagram of another. The bridge of four diodes is packaged in a W500M case, which has four leads. The lead identified with a + sign

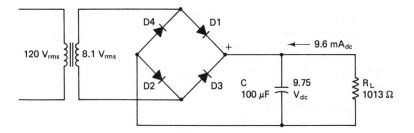

Figure 2-7 Full-wave bridge rectifier without three-terminal regulator shown in Figure 1-1(a).

is connected to the filter capacitor. The one opposite to it goes to the low side of the circuit. The other two leads are connected to opposite ends of the transformer secondary winding.

The bridge conducts when V_{sec} is greater than V_c. When the upper end of the secondary is positive and the lower end negative, $D1$ and $D2$ are forward biased and conduct, but $D3$ and $D4$ are reverse biased and do not conduct. When the upper end of the secondary is negative and the lower end positive $D3$ and $D4$ are forward biased and conduct, but $D1$ and $D2$ are reverse biased and do not conduct. In either case, on each peak the circuit current flows in the same direction from the junction of $D2$ and $D4$, through R_L, to the junction of $D1$ and $D3$.

A diode bridge of this type typically has a *PIV* of 50 or 100 V_{pk} and a maximum I_F of 1.4 A. The diodes do not have to withstand a *PIV* any higher than the single diode in a half-wave rectifier, and they use the same type of transformer.

In our exercise, V_{sec} was measured by the DMM as 8.1 V_{rms}. The peak value of this is 11.5 V_{pk}, and V_c was measured as 9.75 V_{dc}. The measured value of R_L was 1013 Ω, so I_L was 9.75/1013 = 9.6 mA$_{dc}$. As explained previously, t now is 5 ms. Using Equation (2-1) gives

$$V_r = \frac{9.6 \times 10^{-3} \times 5 \times 10^{-3}}{100 \times 10^{-6}}$$

$$= 0.48 \; V_{pp}$$

$$= 0.2 \; V_{rms}$$

The DMM confirmed this by measuring 0.2 V_{rms} across the filter capacitor.

The ripple factor is given by V_r/V_{dc} and is also given by Equation (2-2):

$$r = \frac{\sqrt{2}}{6.28 \times 120 \times 1013 \times 100 \times 10^{-6}}$$

$$= 0.02$$

Since $r = V_r/V_{dc} = 0.2/9.75 = 0.02$, you can see that we have another confirmation of the value of V_r.

Question: How did the ripple factor r in the circuit in Figure 2-7 compare with r in the circuit in Figure 2-5? [It was half the value.]

CALCULATING THE VALUE OF THE FILTER CAPACITOR

Using Equation (2-3),

$$C = \frac{\sqrt{2}}{6.28 \times 120 \times 1013 \times 0.02}$$

$$= 92.6 \cong 100 \ \mu\text{F}$$

FULL-WAVE CENTER-TAP RECTIFIER

Figure 2-8 shows another type of full-wave rectifier. This uses a center-tapped transformer and two diodes, which conduct in turns as V_{sec} reverses polarity. It has two disadvantages compared with the bridge-type rectifier. The transformer is not so simple, and since V_{sec} appears across each half of the secondary, the reverse voltage across the nonconducting diode is $2\ V_{\text{sec}}$. In view of the fact that diode bridges are very inexpensive, the bridge-type rectifier is generally preferred. In other respects, the full-wave center-tap rectifier power supply behaves in the same way as the full-wave bridge rectifier power supply.

REGULATION

When we were examining the circuit in Figure 2-5, we found that the average output voltage V_o was 9.6 V_{dc}, resulting in an I_L of 9.7 mA$_{\text{dc}}$. Now, if we remove R_L, I_L obviously becomes 0, and V_o becomes 11.0 V_{dc}. The difference of 1.4 V_{dc} in V_o from no load to full load (1 kΩ) is the *load regulation* (14%).

If we now change the input voltage from 120 V_{rms} to 110 V_{rms} (I_L is still 0), V_o decreases to 9.9 V_{dc}. The difference of 1.1 V_{dc} from 120 V_{rms} to 110 V_{rms} is the *line regulation* (11%).

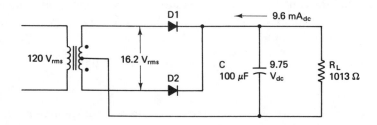

Figure 2-8 Full-wave center-tap rectifier power supply.

Such changes in power supply voltage cannot be tolerated in most applications. Ripple and regulation specifications of less than 100 mV are very common in modern equipment, so an efficient means of regulation is required.

To perform this function, two types of voltage-regulator circuits are used. The first type uses a zener-diode reference alone. The second type uses a combination of zener-diode reference and transistors. Regulators are also available as ICs, which are now widely used.

ZENER DIODES

The zener diode has a very sharp voltage breakdown in the reverse-bias region, as shown in Figure 2-9. The voltage at which this takes place provides the reference. It is called the *breakdown voltage* $V_{Z,\text{nom}}$. Other important parameters are the *minimum current necessary for operation* I_{ZK} (also called the knee current), the *maximum current that can flow in the diode* I_{ZM}, *the test current* I_{ZT}, and the *zener impedance* Z_Z. These are all given in the manufacturer's specification for each diode.

ZENER-DIODE SHUNT REGULATOR

A circuit using a zener-diode shunt regulator is shown in Figure 2-10. It will serve to illustrate the meaning of the zener parameters. The diode in this circuit has the following characteristics:

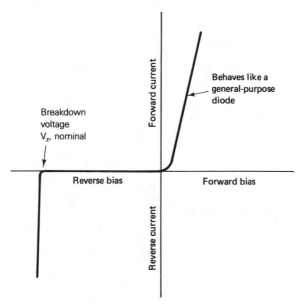

Figure 2-9 Zener diode characteristic curve.

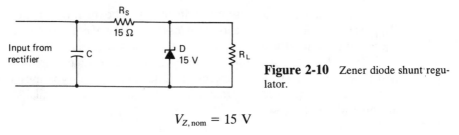

Figure 2-10   Zener diode shunt regulator.

$$V_{Z,\text{nom}} = 15 \text{ V}$$

$$I_{ZT} = 75 \text{ mA}$$

$$Z_Z = 2 \ \Omega$$

This circuit could be used in a low-current application in which the input voltage and load are fairly constant. The zener diode provides constant voltage as long as its current exceeds I_{ZT}. However, there are slight variations depending on its impedance. The output voltage V_o is given by

$$V_o = V_Z = V_c - V_{R_s} = V_c - I_s R_s \qquad (2\text{-}5)$$

$$I_s = I_Z + I_L \qquad (2\text{-}6)$$

$$V_Z = V_{Z,\text{nom}} + (I_Z - I_{ZT})Z_Z \qquad (2\text{-}7)$$

Take the case where V_c varies from 18 to 16.5 V_{dc} and maximum I_L is required to be 100 mA_{dc}. Then, at $V_c = 18 \ V_{\text{dc}}$ and $I_L = 0$,

$$I_Z = I_s = \frac{V_c - V_{R_s}}{R_s}$$

$$= \frac{18 - 15}{15} = 0.2 \text{ A}$$

$$V_Z = 15 + 2(0.2 - 0.075) \qquad (2\text{-}8)$$

$$= 15.25 \ V_{\text{dc}}$$

Now, when $V_c = 18 \ V_{\text{dc}}$ and $I_L = 100 \ \text{mA}_{\text{dc}}$,

$$I_Z = 0.2 - 0.1 = 0.1 \ A_{\text{dc}}$$

$$V_Z = 15 + 2(0.1 - 0.075)$$

$$= 15.05 \ V_{\text{dc}}$$

Therefore, the load regulation is $15.25 - 15.05 = 0.2 \ V_{\text{dc}}$. This is a change of 1.3%.
At $V_c = 16.5 \ V_{\text{dc}}$ and $I_L = 0$,

$$I_Z = \frac{16.5 - 15}{15} = 0.1 \ A_{\text{dc}}$$

$$V_Z = 15 + 2(0.1 - 0.075)$$

$$= 15.05 \ V_{\text{dc}}$$

Therefore, the line regulation is $15.25 - 15.05 = 0.2$ V_{dc}, the same as the load regulation (1.3%).

It is clear from the foregoing example that even a simple zener circuit makes a considerable difference in regulation. It also reduces ripple voltage. Since the change in V_c of 1.5 V_{dc} results in a change in output of only 0.2 V_{dc}, any ripple in the input will also be attenuated by a factor of $1.5/0.2 = 7.5$

> ***Question:*** What values of zeners are available? [All values from 2 to 200 V_{dc}. Commonly used are 5.1, 6.2, 9.1, 12.0, and 15.0 V_{dc}.]

SERIES TRANSISTOR REGULATOR

Figure 2-11 shows a regulator using a transistor. The value of resistor R_s is chosen so that the zener diode current is approximately in the middle of its operating range $(I_{ZM} - I_{ZK})/2$. The transistor must be capable of carrying all the load current and withstanding a voltage equal to $V_c - V_o$, so it is often a power transistor.

In this regulator, when V_o tends to decrease for any reason, V_{be} will increase, with a consequent increase in V_{ce}. The increased current through the transistor and the load causes V_o to remain at the proper value. The opposite happens when V_o tends to increase.

If V_c should tend to increase, V_b would also do the same, resulting in an increase in V_{be}, resulting in an increase in V_{ce}. Since $V_o = V_c - V_{ce}$, the transistor absorbs the increase in V_c, so V_o does not change. The equations for this circuit are

$$V_o = V_c - V_{ce} = V_Z - V_{be} \qquad (2\text{-}9)$$

$$I_s = I_Z + I_b = \frac{V_c - V_Z}{R_s} \qquad (2\text{-}10)$$

$$I_b = \frac{I_L}{h_{FE}}, \qquad \text{where } h_{FE} = \text{dc beta} \qquad (2\text{-}11)$$

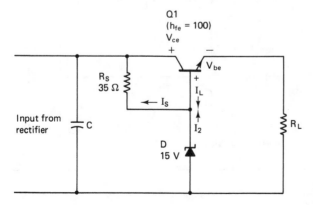

Figure 2-11 Series transistor regulator.

If we were to calculate regulation and ripple factor using the same zener characteristics and $h_{FE} = 100$, as in the previous example, we would find that they are both improved considerably. Furthermore, the circuit is capable of passing a much greater load current.

SERIES REGULATOR WITH FEEDBACK

The foregoing two regulated power supplies have certain disadvantages. Regulation is relatively poor, and V_o cannot be adjusted. The power supply shown in Figure 2-12 overcomes both of these disadvantages by using feedback.

Q_1 and Q_2 form a differential amplifier that compares the voltage on the base of Q_1 (which is a fraction of V_o) to V_{REF}. If V_{b1} is greater than V_{REF}, Q_1 conducts more, decreasing the base voltage of Q_3. This causes Q_3 and Q_5 to conduct less, causing V_o to decrease until $V_{b1} = V_{REF}$. If V_{b1} is less than V_{REF}, Q_1 conducts less, and the base voltage of Q_3 increases. Q_3 and Q_5 therefore conduct more, and V_o increases.

Since V_o is given by

$$V_o = \frac{V_{b1}(R_A + R_B)}{R_B} \qquad (2\text{-}12)$$

any change in R_A results in a change in V_o; and since R_A is adjustable, we can adjust V_o to any value we want, within limits.

Transistor Q_4 limits the output current of the power supply. When the voltage

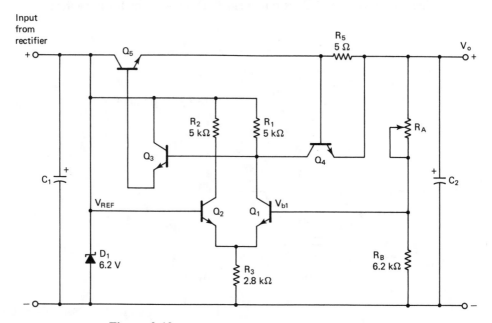

Figure 2-12   Adjustable series regulator with feedback.

across R_5 reaches $0.6\ V_{dc}$, Q_4 conducts, decreasing the base voltage on Q_3. This causes Q_3 and Q_5 to conduct less, causing V_o to decrease.

THREE-TERMINAL REGULATORS

The schematic diagram in Figure 2-13 shows how much more advanced an integrated-circuit regulator is than any we have considered so far. This IC has three terminals: input, output, and common. The fixed output voltage is set by the

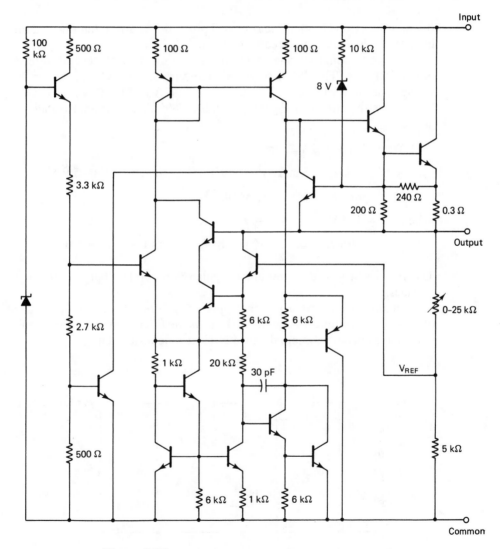

Figure 2-13 Typical internal circuit of three-terminal regulator.

manufacturer's selection of the value of the output resistor. The output voltage can also be varied by the user by connecting two external resistors, as shown in Figure 2-14. Maximum output current with heat sink is 1.5 A or better.

The regulation and ripple rejection characteristics for this regulator are as follows:

> Line regulation, typically 0.1%
> Load regulation, typically 0.1%
> Ripple rejection, −80 dB

In addition, the IC has built-in overload protection, current limiting, thermal overload protection, and safe area protection.

Three-terminal voltage regulators are available with several different fixed output voltages. The 78XX series is widely used. The XX stands for the actual voltage output, for instance, 7805 (5 V), 7812 (12 V), 7815 (15 V), and so on. Figure 2-15 shows the circuit of Figure 2-5 with a 7805 three-terminal voltage regulator added.

When $V_{pri} = 120$ V$_{rms}$ and $V_{sec} = 8.0$ V$_{rms}$,

$$\text{with } I_L = 5 \text{ mA}, \quad V_c = 9.9 \; V_{dc} \qquad and \qquad V_o = 5.0 \; V_{dc}$$

$$\text{with } I_L = 0, \qquad V_c = 10.3 \; V_{dc} \qquad and \qquad V_o = 5.0 \; V_{dc}$$

When $V_{pri} = 110$ V$_{rms}$ and $V_{sec} = 7.3$ V$_{max}$,,

$$\text{with } I_L = 0, \qquad V_c = 7.8 \; V_{dc} \qquad and \qquad V_o = 5.0 \; V_{dc}$$

The 79XX series of three-terminal regulators is similar, but is used for negative voltages.

Figure 2-16 shows a power supply using three three-terminal regulators to obtain three different voltages from the same rectifier. Power supplies with multiple voltage outputs are commonly used in electronic instruments.

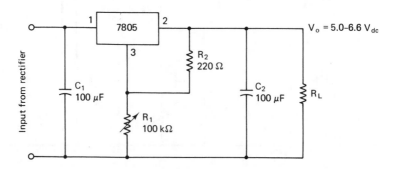

Figure 2-14 Fixed-output regulator made variable by addition of $R1$ and $R2$.

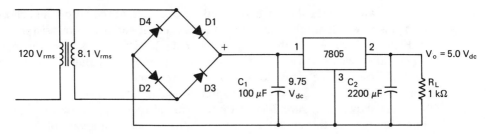

Figure 2-15 Bridge rectifier power supply with three-terminal regulator.

SWITCHING REGULATOR

Until now we have been looking at *linear* regulated power supplies. This term means that as the input voltage increases the input power increases. But since the output power remains the same, there is no place for the additional power to go except in the form of heat in the power supply circuit components. This is very inefficient and gets to be serious in high-power supplies. The *switching regulator* was designed to overcome this.

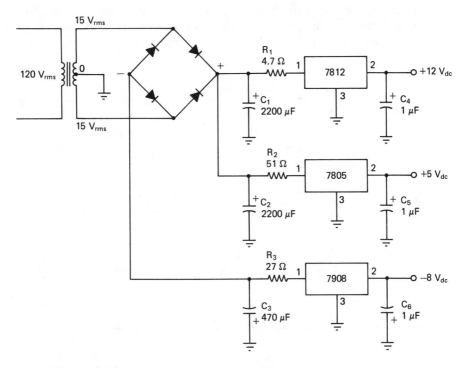

Figure 2-16 Regulated power supply with three outputs. Note how negative voltage output is obtained and how all three supplies are returned to the center tap of the transformer.

A switching regulator works by taking a sample of the output and chopping it into pulses, which are then applied to the base of the series pass transistor, as shown in Figure 2-17. If either the line voltage or the load voltage starts to increase, the pulse duty cycle decreases correspondingly. This makes the pulses narrower so that each pulse turns the transistor on for a shorter time. The pulses appearing at the emitter are converted to dc by the filter choke and capacitor.

As shown in Figure 2-17, a three-terminal regulator is used for a switching power supply. In this circuit a fixed regulator is used, but a variable one could be used instead.

TROUBLESHOOTING POWER SUPPLIES

Power supplies can be tough to troubleshoot, because their ramifications extend into all parts of the equipment they serve. An apparent power supply fault may not be caused by the power supply at all. Conversely, a problem in the power supply may show up in another section altogether.

Figure 2-18 shows a typical power supply as found in a piece of equipment with multiple requirements. In this case it is a cordless telephone. The power supply has to provide voltages to charge the batteries in the handset unit and operate the transmitter, receiver, and various ICs (including a microprocessor) in the base unit. There are three regulators, one for each supply voltage. The three-terminal regula-

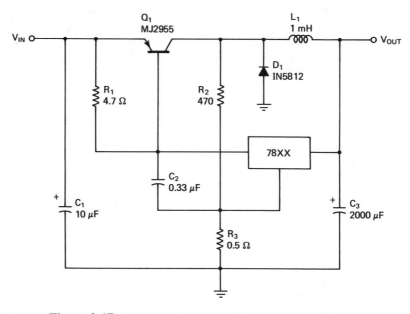

Figure 2-17 Switching regulator using a three-terminal regulator.

tor for the logic circuitry is located alongside the microprocessor, rather than in the power supply itself.

> *Question:*   Why aren't three-terminal regulators used throughout? [The use of a three-terminal regulator depends on whether it can deliver the output current or voltage required.]

Figure 2-18 shows the normal voltages. But let us suppose that something has gone wrong. The equipment does not work. None of the indicator LEDs are lit, not even the "battery low" light. Where is the problem, in the power supply or in one of the other circuits?

Assuming that there is power at the wall outlet and that no internal fuses have blown, we would expect to find some voltage on the transformer secondary, and this we find, and also on the input filter capacitor, though lower than normal. The output of $Q1$, however, which should be 8.2 V_{dc}, is only 1.5 V_{dc}, and the output of $Q4$ is zero. The voltage on its base is also only 0.7 V_{dc}. It looks as if the trouble may be in the second regulator.

Since there is a voltage on $Q4$'s collector, even though of the wrong value, we can assume that C_1 is not shorting it to ground. C_2 could be shorting V_B to ground, but we measured 0.7 V_{dc} on the base, so C_3 becomes our prime suspect. The ceramic 0.047-μF capacitor in parallel with C_3 could also ground the emitter of $Q4$, but ceramic capacitors are much less likely to short than electrolytics, so we'll ignore it for the time being.

To verify that C_3 is indeed grounding $Q4$'s emitter, we unsolder one of C_3's leads and see if this restores the proper V_E. If it does, then we have pinpointed the cause of the problem. (V_E will actually measure about 5.1 V_{dc} when C_3 is disconnected.) When we replace C_3 with a good capacitor of the same value and voltage rating, the cordless telephone should be back in business.

However, suppose disconnecting C_3 does not restore voltage to $Q4$'s emitter? Unfortunately, there are several bypass capacitors in other sections that could be grounding the supply. But now we do know that the problem is not in the power supply. It will be necessary to disconnect other sections, one at a time, until we find where the 5.4-V_{dc} supply is being grounded. Bypassing and decoupling capacitors that supplement the power supply filters are provided in most systems. We look for them at the V_{cc} and ground pins of various driving and receiving devices. Where they are for radio-frequency decoupling they will have values from 0.01 to 0.1 μF. The latter may also be found at the power transformer primary terminals.

At this point we might consider an interesting exercise. We can reproduce this stage of the power supply on our breadboard and supply it with an input of 8.2 V_{dc}. Then we can replace each component, one at a time, with a short piece of hookup wire, simulating an internal short, and see what this does to the voltages on $Q4$. Similarly, we can remove the components, one at a time, to simulate an open, and see what this does. Our results would be pretty much the same as these:

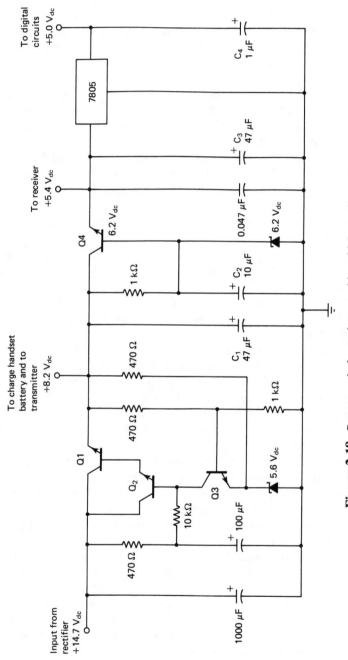

Figure 2-18 Power supply for equipment with multiple requirements.

Component Shorted	Voltages on $Q4$		
	V_C	V_B	V_E
C_1	0	0	0
C_2	7.3	0	4.9
C_3	1.5	0.7	0
Ceramic capacitor	1.5	0.7	0
Zener diode	7.3	0	4.9
$Q4$	8.2	6.2	8.2
1-kΩ resistor (could not short)			
Component open			
C_1	8.2	6.2	5.4
C_2	8.2	6.2	5.4
C_3	8.2	6.2	5.1
Ceramic capacitor	8.2	6.2	5.1
Zener diode	8.2	8.2	7.7
$Q4$	8.2	6.2	0
1-kΩ resistor	8.2	0	4.9

For a quick check, most technicians would use an ohmmeter to verify that, in fact, a suspected capacitor was really shorted and would bridge a capacitor suspected of being open with a similar type to see if that restores normal operation. Open capacitors have the effect of increasing ripple and reducing regulation.

We may have this trouble also when a large filter capacitor has lost *some* of its capacitance or becomes leaky. For example, increased ripple may be causing erratic performance. Such a capacitor should be tested with a capacitance meter. However, if we do not have one, we can easily test it as follows.

Test Procedure   Connect the capacitor in series with a variable resistor, and apply a 60-Hz ac voltage, as shown in Figure 2-19.

In this example, we are testing a 10-μF capacitor. We adjust the potentiometer until the voltage dropped across it is the same as that dropped across the capacitor. Then we disconnect the voltage source and

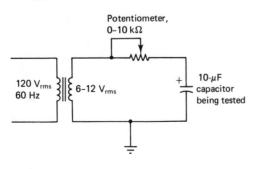

Figure 2-19 Checking the capacitance of a capacitor (see text).

measure the resistance across the "pot." This turns out to be 530 Ω. Since the voltage dropped across this resistance is the same as that dropped across the capacitor, this must also be equal to the reactance (X_c) of the capacitor. The capacitance C of the capacitor is given by

$$C = \frac{1}{2\pi f X_c}$$

$$= \frac{1}{6.28 \times 60 \times 530}$$

$$= 5.0074 \times 10^{-6}$$

$$\cong 5 \ \mu F$$

Obviously, the capacitor has only about 50% of the capacitance it is supposed to have and should be replaced.

3

The Function Generator

Standard function generators offer a variety of waveforms, such as sine, square, triangle, and ramp. Some also have pulse capabilities.

A typical function generator is shown in block form in Figure 3-1. The square wave from an astable multivibrator, or other similar circuit, goes to the upper terminal of the function selector switch, and thence to the output jack. In this case there is an output amplifier, but one is not always required.

At the same time, the square wave is applied to a circuit where it is converted into a triangle wave. This waveform may also be connected to the output.

The circuit used to change the square wave into a triangle wave is called an *integrator*. As shown in Figure 3-2, it uses an op-amp (operational amplifier), with feedback via a capacitor. The positive-going excursion of the square wave starts the capacitor charging with a positive slope, but when the square wave starts its negative-going excursion, the capacitor discharges with a negative slope.

The triangle wave is converted to a sine wave by an op-amp circuit called a *multiplier*. The multiplier uses a diode or a transistor in its feedback circuit. Diodes and transistors have an inherent exponential and therefore logarithmic characteristic; consequently, the output voltage is proportional to the logarithm of the input voltage.

As shown in Figure 3-3, when a triangle wave is applied to the multiplier's input, its output increases exponentially to give a sine wave. When the triangle wave reaches its positive peak, it starts falling, and the sine-wave output does the same. On reaching the zero, the triangle wave continues on down to its negative peak, and the sine output follows suit.

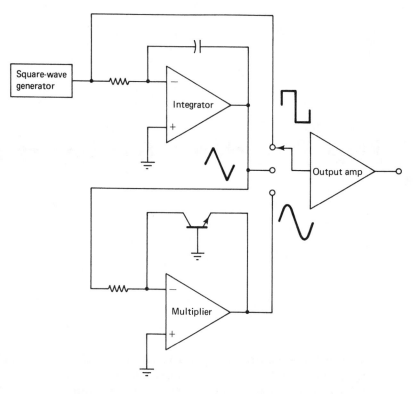

Figure 3-1   Typical function generator block diagram.

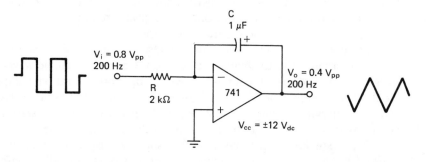

Figure 3-2   Integrator converts square wave to triangle wave.

FUNCTION GENERATOR IC

Figure 3-4 is a block diagram of the XR-2206, which is a widely used medium-scale IC function generator. In addition to having the capability of producing sine, square, and triangle waves over a frequency range from 0.01 Hz to 1 MHz, it can generate pulses with duty cycles from 1% to 99%. Its sine- and triangle-wave outputs can also

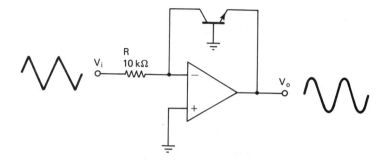

Figure 3-3 Multiplier converts triangle wave to sine wave: $v_o = (-26\text{ mV})(\ln v_i/R - \ln I_{ES})$. I_{ES} is the base–emitter reverse saturation current.

be modulated, both AM and FM, and it can be used for frequency-shift keying (FSK), which is used by modems for transmitting data over telephone lines.

Question: What do we mean by a *medium-scale* IC? [ICs are classified as *small-scale integration* (SSI), with up to 12 gates; *medium-scale integration* (MSI), with from 12 to 100 gates; *large-scale integration* (LSI), with from 100 to 1000 gates; and *very large-scale integration* (VLSI), with over 1000 gates. The XR-2206, therefore, must have between 12 and 100 gates, or components.]

The *voltage-controlled oscillator (VCO)** generates basic periodic signals, which are then converted into the waveforms appearing at the outputs. The frequency is set by an external timing capacitor and resistor. This resistor is part of a voltage divider. The other part is internal. The value of the resistor determines the voltage applied to the VCO.

The external capacitor can be a single capacitor, but to obtain the full range of frequencies that the VCO can generate, several capacitors can be mounted on a switch, as shown in Figure 3-5(a). The same can be done for the external resistor, as in Figure 3-5(b). There must be a fixed 1-kΩ resistor between the potentiometer and the input pin, however, to limit the current to a maximum of 3 mA.

Question: What is the effect on the frequency of using a smaller capacitance? [The smaller the capacitance is the higher the frequency.]

The generator's output is applied to the sections shown in Figure 3-4 for conversion into the desired waveforms.

The square-wave output is obtained by means of a transistor that has an open collector connected to pin 11. The transistor is isolated from the VCO by a buffer amplifier, so its operation will not affect the frequency of the latter. The purpose

*See Chapter 5.

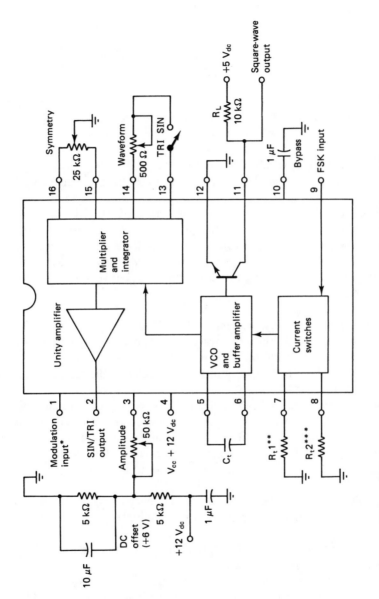

Figure 3-4 Block diagram of function generator IC XR-2206. *, ground if not used. **, minimum value of 1 kΩ. ***, only required for FSK.

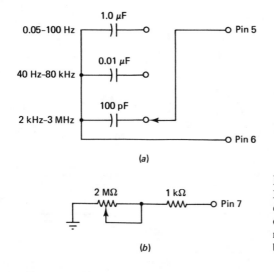

(a)

(b)

Figure 3-5(a) Range switch for XR-2206 function generator. (b) Frequency control for XR-2206 function generator (the same arrangement may be connected to pin 8 also for use when FSK is being used).

of the open collector is to let us connect a dc voltage, via a load resistor, to the collector, so that we can get a desired square-wave voltage. The square-wave voltage will depend on the applied voltage.

The wave-shaping circuit contains both an integrator and a multiplier. An external switch is connected to pin 13. When this switch is open, a triangle wave appears at pin 2. When it is closed, an external harmonic-adjusting potentiometer is connected between pins 13 and 14. This causes the triangle wave to be applied to the multiplier so that a sine wave appears at pin 2. Its symmetry can be fine-tuned by means of a potentiometer connected between pins 15 and 16.

V_{cc} of 12 V_{dc} is applied at pin 4. Pin 3 is used to adjust the amplitude of the sine and triangle waves and to set their offset voltage, but it has no effect on the square wave. The maximum voltage that can be applied to pin 3 is one-half V_{cc}. The actual output voltage depends on the setting of the external potentiometer connected to pin 3.

To demonstrate how this function generator works, we will build the circuit shown in Figure 3-6 on our breadboard. When we apply power to this circuit, we connect our scope to pin 11 first and verify that we have a square wave. This should have an amplitude of 5.0 V_{pp} and a frequency of 8.3 kHz approximately (any value between 8 and 9 kHz is acceptable, since component values may vary).

Now, with the switch at pin 13 open, we transfer the scope probe to pin 2. We should see a triangle wave. We should adjust the potentiometer connected to pin 3 for a maximum waveform without clipping of the peaks. This will be about 6.8 V_{pp}, at the same frequency as the square wave.

Now we close the switch connected to pin 13 and should see a sine wave. It will probably need some improvement. To effect this, we first set the potentiometer connected between pins 15 and 16 to the middle of its range. Then we adjust the shape of the sine wave with the potentiometer connected to pin 14. When we have

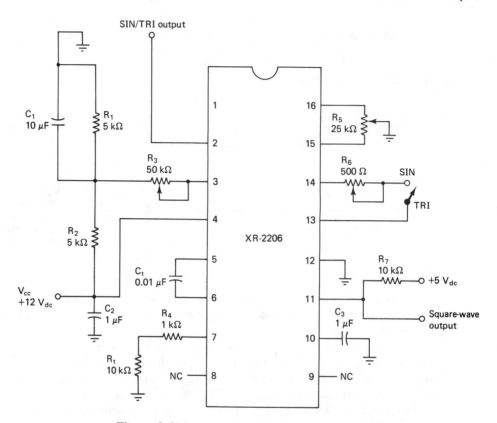

Figure 3-6(a)   Operating function generator XR-2206.

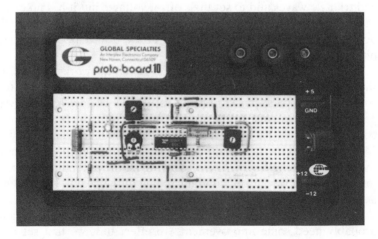

Figure 3-6(b)   Breadboard realization of Figure 3-6(a).

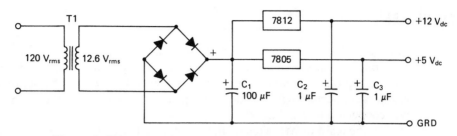

Figure 3-6(c)   Power supply for function generator XR-2206 (if required).

done this, we give it a final touch up with the potentiometer connected between pins 15 and 16.

The sine wave will have an amplitude of about 2.7 V_{pp} and the same frequency as the square wave. The amplitudes of the triangle wave and the sine wave can be adjusted downward with the potentiometer connected to pin 3, but we cannot increase them without distorting the waveform.

> *Question:*   Which has the greater amplitude, the triangle wave or the sine wave? [The triangle wave.]

A very approximate figure for the frequency can be given by

$$f_0 = \frac{1}{RC} \tag{3-1}$$

where R is the resistance connected to pin 7 and C the capacitance connected between pins 5 and 6; but the value measured by a frequency counter, or even by the scope, is much more accurate. For instance, if we calculate the frequency in this case, we get

$$f_0 = \frac{1}{11 \times 10^3 \times 0.01 \times 10^{-6}} = 9.1 \text{ kHz}$$

whereas we already know it is 8.3 kHz.

> *Question:*   Why *is* there a difference between the calculated value and the measured value? [The frequency calculated above is the reciprocal of the time constant RC. But the voltage built up on C after one time constant is really only 63.2% of the applied voltage, because the current charging C decreases as the difference between the voltage on C and the applied voltage gets less. Consequently, the capacitor actually takes longer to charge than the time constant, and therefore the real frequency $(1/RC)$ will be lower.]

PULSE GENERATOR

The XR-2206 can also be used as a pulse generator. For this, the square wave output from pin 11 is modified as shown in Figure 3-7. By connecting pin 11 to pin 9, we are causing the internal current source to switch back and forth between the two resistors connected to pins 7 and 8 (in square-wave generation the timing resistance was connected to pin 7 only).

If the resistances are unequal, the lengths of time the current flows for the high and low states of the pulse are unequal also. This results in an unequal duty cycle, and we can select the percentage by our choice of resistor values. The resistor connected to pin 7 sets the duration for the upper level of the pulse; the resistor connected to pin 8 sets the duration for the lower level. In each case, a minimum resistance of 1 kΩ is necessary so that the current limit of 3 mA is not exceeded.

> *Question:*   What happens if the resistors connected to pins 7 and 8 are equal? [The duty cycle is 50%.]

If we look at the output of pin 2 in this circuit, we find that the sine and triangle waves are distorted in proportion to the duty cycle of the pulse. If the duty cycle percentage is high, enough, the triangle wave becomes a sawtooth, or ramp.

MODULATION

The signal generated by the XR-2206 can also be amplitude modulated by applying a modulating frequency to pin 1. This is done using the circuit shown in Figure 3-8. The sine-wave "carrier" frequency is 83 kHz, and the modulating frequency (derived from the power line via a step-down transformer) is 60 Hz.

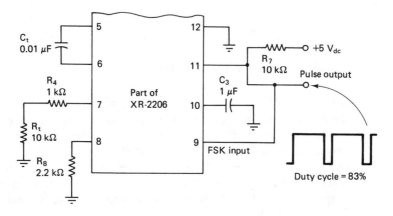

Figure 3-7 By modifying Figure 3-6, as shown above, the XR-2206 becomes a pulse generator.

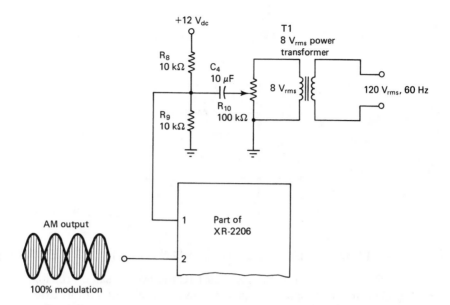

Figure 3-8 Using the modulation input of the XR-2206. The circuit in Figure 3-6 is modified as shown. The percentage of modulation is adjusted by $R10$. C_t is 0.01 μF, R_t is 1 kΩ, to give a carrier of 83 kHz. $R8$ and $R9$ clamp the dc level at 6 V_{dc}.

The potentiometer connected across the modulating signal input enables us to adjust the modulation percentage. Generally, we want 100% modulation, as shown in Figure 3-8.

DC OUTPUT LEVEL

With the circuit connected to pin 3 that we have been using so far, the XR-2206 output from pin 2 consists of a sine or triangle wave centered about a dc voltage equal to 0.5 V_{cc}. However, if a variable dc offset is desired, we can change the circuit connected to pin 3 for the one shown in Figure 3-9. This allows the dc voltage at the input to the 50-kΩ potentiometer to be set between 2.8 and 9.2 V_{dc}, if V_{cc} is 12 V_{dc}.

Question: Why would we want to change the dc offset voltage? [With the circuit shown in Figure 3-6, the dc offset will be approximately the same as the dc bias at pin 3, which is half the source voltage, or 6 V_{dc}. Some circuits will not work, or may even be damaged, if such an offset is used. The circuit in Figure 3-9 allows us to change the offset within the limits given. However, if we want *no* offset at all, we can connect a capacitor of suitable value between the output (pin 2) and the circuit to which we want to apply the function.]

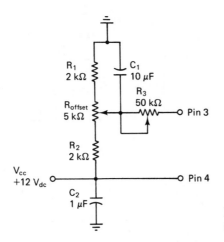

Figure 3-9   Circuit for variable dc off-set.

TROUBLESHOOTING FUNCTION GENERATORS

Troubleshooting a function generator IC follows the same principles as in Chapter 2. On our breadboard we can simulate the effects of various external component failures. If none of these simulations duplicates the symptom of the faulty circuit, then we can assume the IC is to blame.

As an example of the first method, let us rebuild on our breadboard the circuit in Figure 3-6, in which the function generator is being used as a signal source. We will suppose that we have a malfunction such that the square-wave output is normal, but the triangle- and sine-wave outputs are badly distorted, as shown in Figure 3-10. We find we cannot adjust them for correct shape and that their amplitude is much reduced.

Measurement of voltage at pin 3 gives 1.4 V_{dc}, so we turn our suspicions in the direction of the voltage divider that should apply 6 V_{dc} to the potentiometer connected to pin 3. We find that the voltage is, in fact, zero. This would suggest that the capacitor in parallel with the 5.0-kΩ resistor to ground is shorted internally. When we replace the capacitor, we restore normal operation.

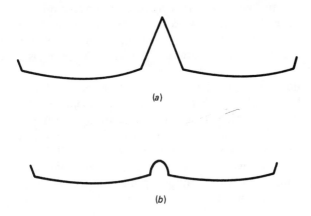

(a)

(b)

Figure 3-10   Badly distorted output waveforms: (a) triangle; (b) sine.

If replacing the capacitor had not eliminated the fault symptoms, we could have continued to test various possibilities in this area. However, we would not have had much success, because nothing other than a short in the capacitor in question would have had the effect that it had. We would have been forced to the conclusion that the IC was to blame.

As a matter of fact, there is very little that could go wrong in this circuit other than the IC, since so much of it is inside the IC. Resistors do not short like capacitors, although they may change value, burn up, or become detached from their leads. So, if we failed to get anything from the output, we would have to assume it was due to a capacitor, and if the square wave was affected, it would have to be the capacitor connected to pins 5 and 6 or the one connected to pin 10. The capacitor connected to pins 5 and 6, if open, would probably still give us a distorted output, but above the upper frequency limit of the function generator (1 MHz).

The capacitor connected to pin 10 would show very little sign of being open (perhaps a slight decrease in amplitude at higher frequencies), but both capacitors would kill the square wave if they were shorted. A defective resistor connected to pin 7 or 8 could also kill the square wave or cause the frequency to be incorrect.

4

Optoelectronic Devices

The ability to convey information in visible form has been greatly enhanced by the invention of the light-emitting diode (LED) and the liquid-crystal display (LCD), together with the digital ICs that control them. Previously, such information was given mainly by panel lamps (tungsten and neon) and by the cumbersome devices known as numerical readout tubes.

The kind of information that can be given by lamps and tubes is very limited, the most common use of tungsten or neon lamps being as pilot lights to let us know if the equipment is energized. Even this is more often done with LEDs unless the ambient brightness is too high. The greatest advantages that tungsten lamps have are their brightness and their availability in any color. However, low-cost LEDs are now being produced in all standard colors except blue, so this second advantage is much less than it was.

LIGHT-EMITTING DIODES

The most commonly used LED is made by growing an *n*-type layer of gallium arsenide phosphide on a substrate of gallium arsenide. A superimposed *p*-type layer is then formed by diffusion into the *n*-type layer. When the resulting junction is forward biased, red light is radiated uniformly in all directions; but only the small fraction of this light that strikes the top surface of the diode can escape. For this reason, the light is emitted mostly along the forward axis. If the LED has a clear lens, the light will emanate from an intense pinpoint; if the lens is a diffused type, the light will be spread over a larger area.

LEDs are low-current devices, with maximum forward currents ($I_{F\max}$) generally ranging from 15 to 100 mA. Forward voltages (V_F) vary from 1.75 to 3.0 V, and reverse (breakdown) voltages (V_R) from 3.0 to 5.0 V. This low V_R is in marked contrast to the *PIV* of rectifier diodes, so an LED must not be exposed to a voltage of the wrong polarity.

Since an LED is a junction diode, its resistance drops to a very low value once it begins to conduct, so a current-limiting resistor has to be connected in series with it for its protection. The value of this resistor is given by

$$R = \frac{V}{I} \qquad\qquad (4\text{-}1)$$

where R is the resistance in ohms, V the source voltage less V_F, and I the I_F (not to exceed $I_{F\max}$). For instance, an LED with $V_F = 2.0$ V and $I_F = 10$ mA, when used with a 5-V supply, should have a resistor as follows:

$$R = \frac{5.0 - 2.0}{0.010}$$

$$= 300\ \Omega$$

Many multimeters cannot check LEDs, because measurement of LED forward resistance typically requires 2.1 V. You can determine the output voltage of your multimeter when it is in the ohms mode with another multimeter in the dc volts mode. Some multimeters have dual ohms ranges, with a lower output voltage for the lowest range. A feature often found is a *diode check* function, which works with ordinary diodes. However, it will probably indicate a good LED to be open.

LEDs IN DIGITAL SYSTEMS

So far we have considered LEDs in terms of voltage, current, and resistance, Ohm's law parameters. Circuits in which these parameters predominate are often loosely called "analog" to distinguish them from "digital," in which the essential parameters are *binary digits* or *bits*.

BINARY NUMBER CODE

As in the case of pilot lights, we see that an LED can say something, even if it is not very much. A combination of LEDs, in various states, can say quite a bit more. Four LEDs hooked up to four switches, as shown in Figure 4-1, can actually say 16 things. This is more than enough to represent all the decimal numbers from 0 to 9.

We do this by making the first LED from the right represent the number 1, the second the number 2, the third the number 4, and the fourth the number 8. Table 4-1 shows how, by energizing different combinations of the four LEDs, we can show all ten numbers from 0 through 9. Since each LED has just two possible conditions,

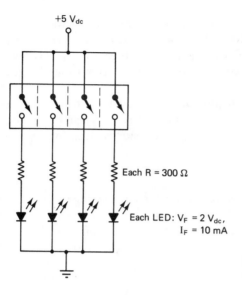

Figure 4-1 Four LEDs can say 16 things.

lit and not lit, this code, which is used universally in computers, calculators, and the like, is called the binary number code. Actually, it uses only 10 of the 16 possible combinations. There are some other codes that use all 16, but we are not concerned with them at the moment.

Question: How many things can eight LEDs say? [2^8, or 256 things; 16 LEDs can say 2^{16}, or 65,536 things; and so on.]

The code in Table 4-1 allows us to send any number from 0 to 9 along four wires just by energizing them with switches. In most cases the switches are transistor

TABLE 4-1 BINARY NUMBER CODE

Binary Number 1 = Lit 0 = Unlit	Value of Each LED When Lit		Total Decimal Value
0 0 0 0	0 + 0 + 0 + 0	=	0
0 0 0 1	0 + 0 + 0 + 1	=	1
0 0 1 0	0 + 0 + 2 + 0	=	2
0 0 1 1	0 + 0 + 2 + 1	=	3
0 1 0 0	0 + 4 + 0 + 0	=	4
0 1 0 1	0 + 4 + 0 + 1	=	5
0 1 1 0	0 + 4 + 2 + 0	=	6
0 1 1 1	0 + 4 + 2 + 1	=	7
1 0 0 0	8 + 0 + 0 + 0	=	8
1 0 0 1	8 + 0 + 0 + 1	=	9

switches, which operate far faster than mechanical switches and are themselves turned on and off by voltages coming to them along other wires.

However, before we can do this there remains the necessity of converting the number we want to send into the binary code and then converting it back into decimal form at its destination. For this, we have to have two integrated circuits, called the *encoder* and the *decoder*.

ENCODER

If we are using a computer or a calculator, we enter the number via a keyboard. On this keyboard there is a key for each of the 10 decimal digits, plus a few more that we shall ignore at present. The keyboard has a circuit called a scan generator that momentarily connects each key in turn to the encoder. It does this continually, at a very high rate, so that any key we touch is connected several times to the encoder, however quickly we do it.

In the encoder (Figure 4-2) are several special switches called *gates*. Gates are explained in Chapter 6, but we can anticipate by saying that the NAND gate used here, and shown also in Figure 4-3, is a device that has a "high" (positive) voltage at its output at all times except when both its inputs are high. When this happens, its output switches to "low" (zero). The actual voltages need not be specified. A high is any value over 2 V (and not over 5 V), and a low is any value below 0.8 V. The high corresponds to the binary 1 and the low to the binary 0.

The other symbol in the diagram, shown again in Figure 4-4, is an inverter. This has only one input and one output. If a high appears on its input, its output goes low, and if a low appears on its input, its output goes high.

Each of the 10 NAND gates in the top row has one of its two inputs connected to a *scan line*, which is also connected to a number key. This input goes high every time the number key is scanned. The other input is connected to all the number keys. If no key is depressed, no voltage appears on this line, and each NAND gate's inputs are both low, except when it is scanned, so its output is high. When a number is scanned, the input of the NAND connected to it goes high. However, the NAND's output remains high, since both inputs have to go high for the NAND's output to go low.

But when a number key is depressed, a high appears on both inputs of the NAND connected to that key when it is scanned, so its output goes low. The encoder now knows that a number (5 in Figure 4-2) has been selected, and it proceeds to turn it into a binary number.

To do this, it uses the four NANDs at the lower right of the figure. Three of these have more than two inputs, but all of the inputs have to be high for the output to go low. The outputs of the 10 NANDs in the top row are all high except for the one connected to key number 5, so we can see which inputs of the lower four NANDs will be high and which low. All the inputs of the NANDs connected to wires 2 and 8 are high, so these wires will be low; but wires 1 and 4 are high because their NANDs

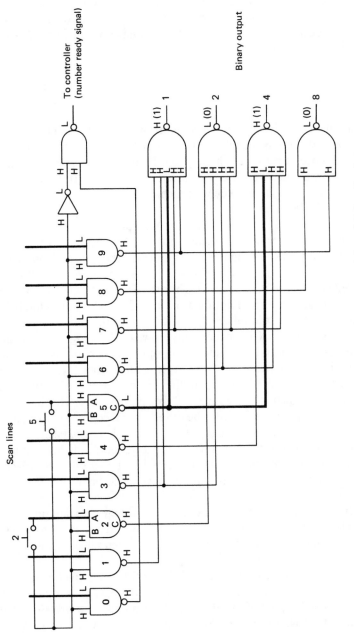

Figure 4-2 Encoder. On the keyboard, key 5 is pressed down (all the other keys are up, as shown for key 2). When the scan voltage appears on scan line 5, it makes the A input to NAND 5 high, and also the B input, so output C goes low. By contrast, when the scan voltage appears on all the other scan lines, their input As do not go high, because their keys were not depressed. Therefore, only NAND 5's output goes low.

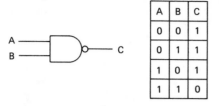

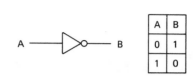

Figure 4-3 NAND gate symbol and truth table that summarizes its operation.

Figure 4-4 Inverter symbol and truth table that summarizes its operation.

do not have all their inputs high. In this way, the binary number 0 1 0 1 has been put on the four wires.

In computers, calculators, and the like, the number we key in will be operated on in various ways and then a result will be ready to be displayed. We shall not worry about the in-between part here but go directly to the end result. This will involve converting binary numbers to decimal numerals by means of a decoder.

DECODER

A decoder, as shown in Figure 4-5, is really very like an encoder working in reverse. The four wires 1, 2, 4, and 8 are shown at the top-left corner. Connected to each is an inverter, so the four single wires become four pairs of wires, the two wires in each pair having signals with opposite polarity. For instance, if external wire 1 is low, one of its two internal wires will be low and the other high.

Assuming the external wires are carrying the signal 0 1 0 1, the inputs of the upper row of NAND gates will be high or low as shown. Their outputs will all be high except that of the sixth from the left. Since all three of its inputs are high, its output is low.

The effect on the NANDs in the bottom-right corner is that they have all inputs high except b and e. Therefore, all outputs are low except b and e. The seven wires from these outputs are connected to a seven-segment numeric display.

SEVEN-SEGMENT NUMERIC DISPLAY

A seven-segment numeric display is an array of seven LEDs. Each LED has a diffused, bar-shaped lens. The segments are arranged in the form of a figure 8, as shown in Figure 4-6, so that any of the numerals 0 through 9 may be displayed by turning on the appropriate segments, always lettered as shown.

There are two types of display. In one, all the anodes of the LEDs are connected together; in the other, all the cathodes. In the common-anode type, the anodes must be connected to the 5-V supply. In the common-cathode type, the cathodes must be connected to ground.

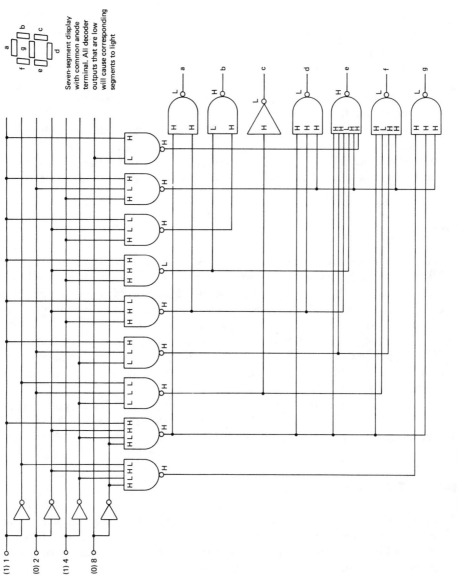

Figure 4-5 Decoder. Binary number 0101 at upper left is decoded to display a 5 on the seven-segment display. Seven-segment display with common anode terminal. All decoder outputs that are low will cause corresponding segments to light.

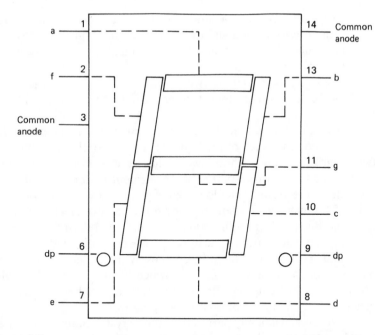

Figure 4-6 Seven-segment display with common anode (available at either pin 3 or 14).

The decoder in Figure 4-5 requires a display with a common anode. Its outputs are connected to the LED cathodes. When an output is low, current will flow from the positive supply through the LED to the low output. When the output is high, no current will flow, or not enough to light the LED.

In this example, the cathodes of segments a, c, d, f, and g are connected to low outputs, so they will light, forming the figure 5. This is the same as the binary number 0 1 0 1, of course, but much more convenient for us.

Decoders are also available in which the outputs to the segments to be lit are high. In such cases, we would use a common-cathode display, with the common cathode grounded. The choice depends on circuit considerations. One type of decoder requires a low input signal wherever a high input appears in the decoder in Figure 4-5.

LIQUID-CRYSTAL DISPLAY

Most readouts for battery-operated devices such as pocket calculators, digital multimeters, and wristwatches are not LED displays, but liquid-crystal displays (LCDs), because the latter use far less current.

Question: What *is* a liquid crystal? [A substance that has both the properties of a solid and the properties of a liquid. That is, although its molecules lie in a certain direction when at rest, as if they were in a solid crystal, the presence of an electric field causes them to turn and take up a new direction, as if they were free to change position, as in a liquid. The molecules revert to their former alignment when the field is removed.]

An LCD does not emit light; it reflects it. As shown in Figure 4-7, it consists of a Dagwood sandwich of the liquid-crystal material between two glass plates and two polarizers. Beneath the lower polarizer is a mirror. The upper glass plate is photoetched with a transparent seven-segment pattern of indium-tin oxide that is invisible to the viewer. The lower glass plate is covered uniformly with indium-tin oxide. This material is conductive and forms the two electrodes of the liquid crystal cell. The thickness of the liquid-crystal film is between 6 and 25 μm.

Light passing through the upper polarizer is polarized. As it continues through the liquid crystal, it is rotated 90 degrees by the molecules of this material. It then passes through the lower polarizer, which is aligned at right angles to the upper polarizer. After being reflected by the mirror, it follows the same path in reverse order until it emerges from the upper polarizer.

However, when any part of the liquid-crystal material is excited by an electric field between the indium-tin oxide electrodes on the upper and lower glass plates, the molecules in that part change position and block the passage of light. The viewer now sees a black numeral. (In some cases we will see a white numeral on a black background, if the polarizers have been arranged that way, but black numerals are more usual.)

An LCD is activated by a square wave with a frequency between 30 and 200 Hz. If dc were used, as in an LED display, the LCD would not clear fast enough after removing the voltage. The frequency is not critical, but if it is lower than 30 Hz, a flicker will be evident in the display. Too high a frequency, on the other hand, will not allow enough time for the liquid-crystal molecules to realign themselves between cycles.

The supporting circuitry is shown in Figure 4-8. Decoding is the same as in LED

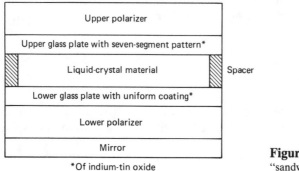
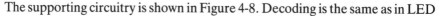
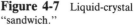

Figure 4-7   Liquid-crystal "sandwich."

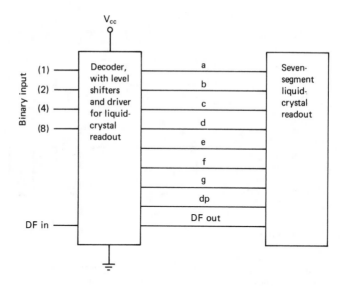

Figure 4-8   LCD and decoder/driver.

displays. The square wave is applied to the DF IN terminal (DF = display frequency). Its amplitude is not critical, because it will be adjusted by the level shifters. The DF OUT signal is applied to the common electrode. Square-wave signals from the display driver that are applied to the segments to be energized have a polarity opposite to the DF OUT signal, so the voltage across the display is doubled. Unactivated segments, on the other hand, are supplied with an in-phase square wave, so the effective voltage across these is zero (see Figure 4-9).

When LCDs are used for television displays, they do not require the high voltages of a CRT. The screen is a matrix consisting of two sets of invisible, conductive parallel lines at right angles to each other, separated by the liquid crystal. The screen is normally dark, but when a line in each set is energized, the voltage difference between them at their point of intersection (a pixel) causes it to become visible, its brightness corresponding to the amplitude of the corresponding picture information. The lines are addressed by vertical and horizontal scan generators synchronized by the sync pulses of the composite television signal. Similar displays are also used for word processors and lap computers.

LCDs require a supply from 3 to 10 V, and consume from 0.3 to 100 mW of power per pixel.

OTHER TYPES OF DISPLAY

The vacuum-fluorescent display is a vacuum tube in which the anode consists of seven segments like the previous displays. They are coated with a fluorescent substance that glows when struck by electrons emitted by the cathode. The segments

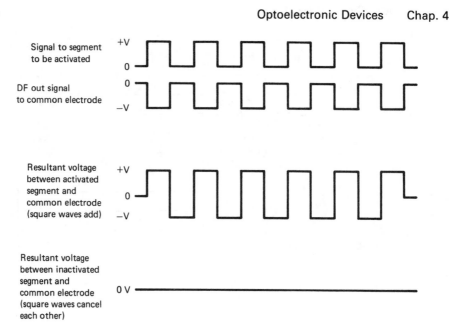

Figure 4-9   Square-wave signals that drive the LCD.

require a positive voltage of from 12 to 60 V_{dc}, so they cannot be excited directly from a decoder, such as those described above. However, this is a bright display and is popular for use in office equipment, microwave ovens, VCRs, and the like, which do not rely on batteries alone.

Cathode-ray tubes are also vacuum tubes, but they require much more supporting circuitry than LEDs or LCDs. They also must have an anode potential of several thousand volts. Apart from television receivers, CRTs are mostly used for computer monitors and oscilloscopes. Since their circuits work on entirely different principles, they will not be discussed here.

DISPLAYS WITH MORE THAN ONE CHARACTER

As you are well aware, pocket calculators and digital multimeters would not be very useful if they could only display single characters. In fact, calculators frequently show as many as 10. Does this mean that there must be a decoder for each character displayed?

In a multiple-character display, the common terminal in each character block (see Figure 4-10) is connected to a scan line from the scan generator. The segment lines are connected in parallel to all the segments of all the character blocks, but only those whose scan line is energized can light. After lighting each character, the scan

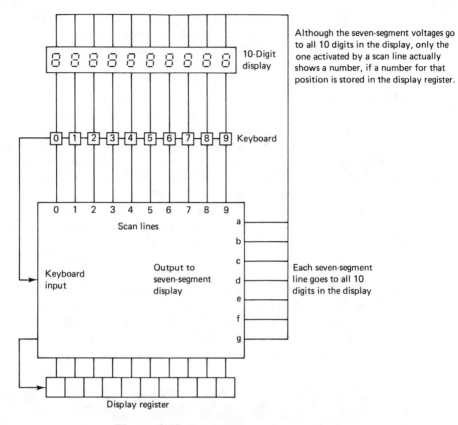

Figure 4-10 Displaying more than one character.

line moves to the next block, and the segments energized in that block light. The blocks are scanned over and over at a high rate, so the eye sees no flicker at all.

OPTICAL ISOLATOR

An optical isolator or optocoupler is a very small four-terminal device that can perform as a solid-state transformer or relay, since an input signal causes an output signal without any electrical connection between the input and output terminals. This is a particularly valuable characteristic where two circuits have a large voltage difference between them, and yet it is necessary to transfer a small signal between them without having to change the voltage level of either.

An optical isolator consists of an optical emitter and an optical detector arranged so that the input signal causes radiation from the emitter, which in turn excites the detector, and switches it from off to on for the duration of the input signal.

The emitter is most often a gallium arsenide LED that emits in the red or infrared region of the spectrum. A typical wavelength is 850 nm (nanometers). This wavelength is particularly suitable for silicon photodetectors, and an optical isolator containing these elements is the most efficient.

Incandescent and gas-discharge (neon) lamps are also used in some optical isolators, but these are designed for special purposes only; and the optical detector may be a photocell, which is much slower than a photodetector. They are not as efficient as the solid-state device.

In the LED–silicon photodetector combination, the detector is most often a phototransistor. The light from the LED causes the phototransistor current to vary as a function of the intensity of the light falling on it. If the input signal current varies, the output signal current will vary accordingly. When used as a switch, an input current turns the detector on; the cessation of the current turns the detector off.

Optical isolators are very fast, with the output current in some changing in as little as 10 to 20 ns (nanoseconds) after a change in the input current. This makes them especially valuable in internal computer circuitry.

In spite of their small size, these devices can provide ac line isolation or isolate circuits that differ by as much as 5000 volts, and because the photodetector can be an SCR or triac, a small input current can control a very large output. In this way a computer can control heavy-duty equipment by means of its own digital signals.

Figure 4-11 shows a breadboard circuit in which an optical isolator, when energized, switches on the lower circuit. There is no signal connection between the two circuits.

Figure 4-11   Optocoupler couples two circuits by light alone. There is complete electrical isolation.

TROUBLESHOOTING OPTOELECTRONIC DEVICES

Testing LEDs

As mentioned earlier in this chapter, many multimeters cannot test an LED as they can other diodes. The simplest way of testing one, of course, is to connect it in series with a current-limiting resistor, apply suitable voltage, and see if it lights.

But this can be a bit risky, since the maximum forward current of some LEDs can be as low as 15 mA, while their maximum reverse voltage is 4 V or less. If we know our LED, we are on safe ground, since we can calculate the proper value for the current-limiting resistance. If not, we can connect a 5-kΩ potentiometer, set to maximum resistance, in series with it, and gradually decrease the resistance until the LED lights.*

However, there is another way, if we have an oscilloscope available. Using our breadboard, we connect a 4.7-kΩ resistor in series with the LED across the secondary of a small power transformer connected to the power line, with the scope probes connected as shown in Figure 4-12.

> *Question:*   What current is the LED subjected to? [About 1 mA$_{rms}$.]

Operate the scope in the X–Y mode (with dc input). If the LED is good, an L shape will appear on the screen (the L may be upside down, depending on which way round the diode is connected). If the LED is open, only a horizontal line will appear. If it is shorted, only a vertical line will appear.

The advantage of this method is that there is no danger of burning up the LED, and it can be connected with either polarity. It may glow faintly, but this is immaterial.

> *Question:*   Can this test be used for ordinary diodes? [Yes. It works for any junction, including those in transistors.]

* Obviously, this doesn't work with an infrared LED. If you are testing one, use the method described next.

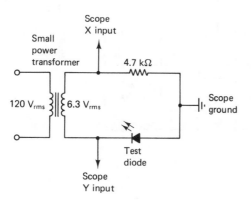

Figure 4-12 Testing an LED with a scope.

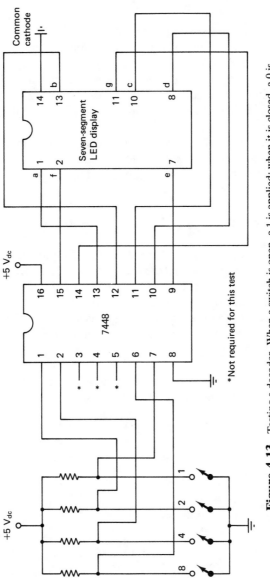

Figure 4-13 Testing a decoder. When a switch is open, a 1 is applied; when it is closed, a 0 is applied. Resistors should each be 4.7 kΩ.

TABLE 4-2   TRUTH TABLE FOR 7448 DECODER

(8) 6	(4) 2	(2) 1	(1) 7		(f) 15	(g) 14	(a) 13	(b) 12	(c) 11	(d) 10	(e) 9	Display (if used)
L	L	L	L		H	L	H	H	H	H	H	0
L	L	L	H		L	L	L	H	H	L	L	1
L	L	H	L		L	H	H	H	L	H	H	2
L	L	H	H		L	H	H	H	H	H	L	3
L	H	L	L		H	H	L	H	H	L	L	4
L	H	L	H		H	H	H	L	H	H	L	5
L	H	H	L		H	H	L	L	H	H	H	6
L	H	H	H		L	L	H	H	H	L	L	7
H	L	L	L		H	H	H	H	H	H	H	8
H	L	L	H		H	H	H	H	H	L	L	9

Column groups: Input Pins[a] spanning the first four columns, Output Pins spanning columns (f)–(e).

[a] Ground input pins where L is shown; otherwise, 5 V.

Seven-segment Numerical Display

You can test a seven-segment numerical readout by mounting it on your breadboard, with the common cathode or anode connected to ground or to the 5-V supply as appropriate, and the segment pins connected via the correct current-limiting resistors to the 5-V supply or ground, as the case may be; or you can use the oscilloscope method just described if you prefer.

Decoder

You can check out a decoder easily with your breadboard. Figure 4-13 shows a 7448 BCD* to seven-segment decoder/driver connected to a 4-bit DIP switch (however, you do not have to have the switch; it's just more convenient if one is around). It is not necessary to have a readout, but it helps. As you can see in the figure, the four switches are used to ground the input pins of the 7448.

Grounding one of its pins is the same as applying a logic low. Applying inputs as shown in Table 4-2, you should get the outputs shown; the figures in parentheses are the values of each input. The lowercase letters refer to the segments as shown in Figure 4-13. The best instrument to use for this type of test is a *logic probe* (see Chapter 6).

* Binary-coded decimal.

5

Timing Pulse Generators: Clocks

In most digital systems there are arrays of flip-flops called registers that are constantly exchanging data with each other and other sections. Utter chaos would result if their operations were not precisely synchronized by clock signals generated by timing pulse generators. In fact, many systems use more than one clock.

In this chapter we are going to look at two ICs that are commonly used as timing pulse generators. They are the quad two-input NAND gate (7400) and the timer (555).

QUAD TWO-INPUT NAND GATE (7400)

In Chapter 4 we showed that NAND gates could have two or more inputs. The output was always high unless all inputs were high, in which case it was low. In the 7400 IC, there are four NAND gates, which are independent of each other, except for sharing the V_{cc} pin (14) and ground pin (7). As we see in Figure 5-1, each of these NAND gates has two inputs only.

The simplest circuit that will give timing pulses is that in Figure 5-2. Each NAND gate has its inputs tied together, so they will be the same. Therefore, if both inputs are high, the output will be low, and vice versa. In fact, we have changed the NAND gate into an inverter.

> *Question:* Couldn't we do the same with two inverters (or two transistors connected as inverters) instead of using NAND gates? [Of course. Wouldn't that be an astable multivibrator?]

60

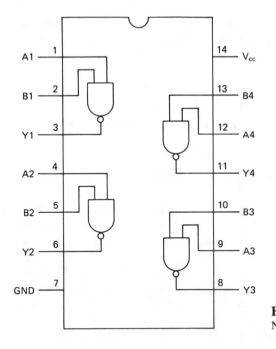

Figure 5-1 7400 Quad two-input NAND gate TTL IC (block diagram).

When power is first applied, the input of the right NAND gate will be low, and its output high. This charges the capacitor to the maximum voltage. However, the charge on the capacitor plate connected to the input of the left NAND gate leaks away through $R1$, so this input, which was high at first, goes low. This causes the output of this gate to go high and, of course, the input to the other gate as well. So its output goes low.

Meanwhile, the high on the input of the right NAND gate leaks away through both resistors rather more quickly, so it becomes a low, the output becomes a high again, and the cycle repeats.

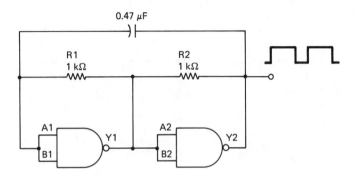

Figure 5-2 Two of the NANDs in a 7400 IC are used here to make a pulse generator. Output waveform has a duty cycle of about 64%.

The time taken by the charge to leak away from the capacitor so that the input to the left NAND gate goes from a high to a low is determined by the value of $R1$ and C, as in

$$T = 2R1\,C \tag{5-1}$$

To take an example, if $R1$ is 1 kΩ and C is 0.47 μF,

$$T = 2000 \times 0.47 \times 10^{-6}$$

$$= 0.94 \text{ ms}$$

The pulse repetition rate (PRR) is the reciprocal of this, or 1 kHz approximately.

Question: Is there any reason for saying *pulse repetition rate* instead of *frequency of operation*? [Not really. Both terms are used interchangeably. However, in *digital* circuits, we use the term PRR as a general rule.]

This generator does not put out a symmetrical pulse. The duty cycle is about 64%. If a symmetrical pulse is required, we must consider another circuit.

Such a circuit is shown in Figure 5-3. We are still using the 7400 IC, but there are *two* timing capacitors. The values chosen for these set the pulse rate; the values of $R1$ and $R2$ have to be between 200 and 220 Ω to provide the right bias (0.6 V_{dc}) on the inputs of NAND gates 1 and 2 so that they will turn on when they are supposed to.

When the input of NAND gate 1 is low its output is high, charging $C1$ to about 3 V_{dc}. This also places a high on the input of NAND gate 2, so its output is low.

However, the high on NAND gate 2's input leaks to ground through $R2$ in the time $R2C1$ seconds, so that it is replaced by a low, and NAND gate 2's output goes high. This has the effect of making NAND gate 1's input high, so its output goes low.

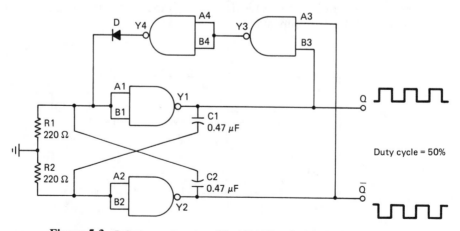

Figure 5-3 Pulse generator using all four NANDs of a 7400 IC. Output waveform is symmetrical if $C1 = C2$.

As we can see, this results in a train of pulses complementary to each other from the outputs of both NAND gates.

> *Question:* What would be the effect if $C1$ and $C2$ had different values? [The pulses and intervals between them would not be equal; in other words, the pulse train would be *asymmetrical*.]

Since there are four NAND gates in the 7400 IC, the other two need not be wasted, but may be connected as shown. This arrangement ensures that the outputs of NAND gates 1 and 2 never go high or low at the same time.

The pulse repetition rate of this circuit is given by

$$\text{PRR} = \frac{0.693}{R2C1} \tag{5-2}$$

For instance, if $R2 = 220\ \Omega$ and $C1 = 0.47\ \mu\text{F}$,

$$\text{PRR} = \frac{0.693}{220 \times 0.47 \times 10^{-6}}$$

$$= 6.7\ \text{kHz} \quad (\text{approximately})$$

The pulse trains are symmetrical, and the circuit will give satisfactory outputs at frequencies below 1 MHz. For higher pulse rates, a clock IC is better. These are crystal controlled and designed to interface with a particular microprocessor. In fact, many are built into the microprocessor chip.

TIMER 555

The 555 IC timer shown in Figure 5-4 consists of two comparators, a flip-flop, a transistor, an inverter, and a voltage divider. The last named provides reference voltages for the comparators that are two-thirds and one-third of V_{cc}.

> *Question:* What is a comparator? [A comparator is an operational amplifier, like the one shown in Figure 4-5. But, whereas op-amps usually have feedback, comparators do not. This is because we want to take advantage of the extremely high gain of an op-amp in the absence of feedback.]

Op-amps have two inputs, inverting ($-$) and noninverting ($+$). With maximum voltage gain, an op-amp's output will swing through its entire dynamic range, V_{min} to V_{max}, as shown in Table 5-1.

In Figure 5-5 a 741 op-amp has a reference voltage V_{REF} of 5.0 V_{dc}. Table 5-2 gives the V_{dc} values measured in this circuit.

In the 555 the V_r for comparator 1 is $2V_{cc}/3$, unless an external voltage is applied

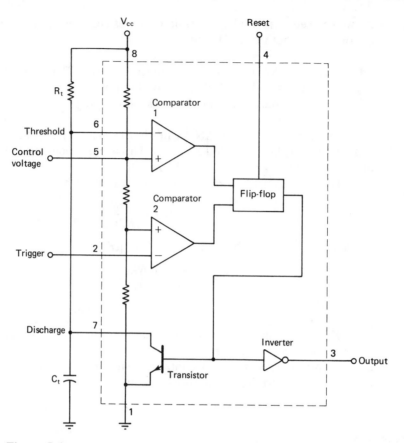

Figure 5-4 Block diagram of 555 timer. Everything within the dashed rectangle is inside the IC.

to pin 5 (control voltage). This pin is connected to the comparator's noninverting input. V_i is derived from the charge on C_t, and when the voltage on this capacitor exceeds V_r, V_o goes low.

V_r for comparator 2 is half the value of that for comparator 1, but in other respects comparator 2's operation is the same as that of comparator 1. The outputs of the comparators are applied to the flip-flop.

Question: What is a flip-flop? [We don't need to know its internal circuitry at this point. It acts like a toggle switch. When either compara-

TABLE 5-1 OP-AMP
COMPARATOR PERFORMANCE

$V_i < V_r$	$V_o = V_{max}$
$V_i > V_r$	$V_o = V_{min}$

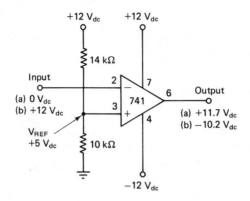

Figure 5-5 Comparator using a 741 op-amp (op-amps are covered in Chapter 9).

tor's output goes high, the flip-flop's input goes high and its output goes low. It stays this way until its input goes low, when its output goes high.]

The output of the flip-flop is applied to the base of the transistor. As long as it is low, the transistor cannot conduct. But when the output of the flip-flop goes high, the transistor turns on, and the charge on C_t is shorted to ground.

The output of the flip-flop is also applied to the inverter, which inverts it so that the voltage on the output pin 3 is the opposite of the voltage on the output of the flip-flop. We can now see how the timer works, but we must first select a capacitor and resistor for C_t and R_t to provide the delay time we want.

We can do this by using the formula

$$t = 1.1 C_t R_t \qquad (5\text{-}3)$$

This is an approximation, but well within the standard values of resistors and capacitors.

For example, if we want a delay time of 10 seconds (s), $C_t R_t$ will be 10/1.1 = 0.91. (In other words, $RC = 0.91$.) We could select a 1.0-μF capacitor and a 9.1-MΩ resistor, which gives $1.1 \times 1.0 \times 10^{-6} \times 9.1 \times 10^6 = 10$ s.

It doesn't matter whether $V_{cc} = 12$ V_{dc}, 5 V_{cc}, or any other value up to the

TABLE 5-2 OP-AMP COMPARATOR PERFORMANCE

V_{in}	V_r	V_o
−12.0	+5.0	+11.7
0	+5.0	+11.7
+2.5	+5.0	+11.7
+5.0	+5.0	+11.7[a]
+12.0	+5.0	−10.2

[a] V_o jumps to V_{min} when V_i exceeds V_r by even the smallest amount.

maximum permissible supply voltage (16 V_{dc}), because the internal voltage divider ensures that the control voltage will be two-thirds of it, and the time taken for C_t to charge to that will therefore be the same.

Before starting, the transistor will be conducting, and the voltage on C_t will be essentially zero. The output voltage will be low also. But as soon as the trigger input is grounded (input low), comparator 2 output will go high, making the input to the flip-flop high also. This causes the latter's output to go low, and this low voltage appears on the base of the transistor, so it ceases to conduct. The output from pin 3 is now high. If we were to connect an LED to it, it would light.

As the transistor is not conducting, C_t starts charging toward V_{cc} via R_t. Eventually, it reaches the control voltage, at which time comparator 1's output goes low, and the low voltage is applied to the flip-flop. This, in turn, switches to a high output voltage, which causes the transistor to conduct, grounding the voltage on C_t. The output on pin 3 now goes low, so an LED, if connected to it, would go out.

We must not forget that the formula for calculating the RC time constant required for obtaining a particular delay applies only when using the 555's internal reference. Since we can use other control voltages, we could use a different value. In that case it would be easier to try various combinations of R_t and C_t on the breadboard rather than calculate, since the calculations are rather difficult. The applicable formula, if you want to do it, is

$$\frac{V_r}{V_{cc}} = (1 - \epsilon^{-t/RC}) \tag{5-4}$$

where ϵ = 2.71828, the base of natural logarithms

$\quad V_r$ = reference voltage

$\quad t$ = delay time

$\quad RC$ = time constant

$\quad V_{cc}$ = supply voltage

Question: But we don't want a timer. What about a clock? [We're just coming to that. The 555 timer can be used either way.]

TIMER 555 CLOCK

Now that we know how the 555 works, we shall have no difficulty in understanding how it may be used as a clock. Figure 5-6 shows the IC connected for this purpose. We now have *two* external resistors in series with the timing capacitor, which charges through $R1$ and $R2$ until the voltage on pin 6 exceeds $\frac{2}{3}$ V_{cc}. Comparator 1's output then goes low, the flip-flop's output goes high, and the transistor conducts.

The capacitor now discharges through $R2$ and the transistor until the voltage on pin 2 drops to $\frac{1}{3}$ V_{cc}. Comparator 2's output then goes high, the flip-flop's output

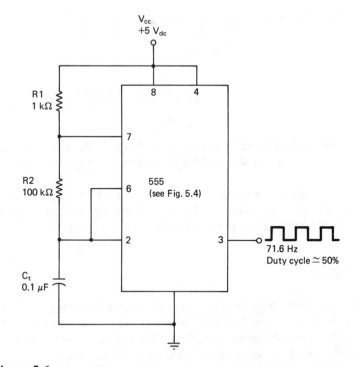

Figure 5-6 Timer 555 connected as a pulse generator. If $R2$ is very much larger than $R1$, the duty cycle will be approximately 50%; otherwise, it will depend on the proportion $R2/(R1 + R2)$.

goes low, and the transistor stops conducting. The capacitor starts charging again, and the cycle repeats, as described above.

Each time there is a low voltage on the base of the transistor the inverter causes a high voltage to appear at pin 3. Conversely, when there is a high voltage on the base of the transistor, a low voltage appears on pin 3. The result is a train of pulses with a pulse repetition rate (PRR) that is determined by the values of $R1$, $R2$, and C_t.

Question: Can the 555 produce pulse trains with any PRR? [Unfortunately, no. Its maximum PRR is about 1 MHz.]

One other thing, you have probably noticed that C_t charges through both resistors, but discharges through only one ($R2$). This means that the discharge time is less than the charge time, so the output pulses cannot have a 50% duty cycle. Therefore, if we want one, we must consider going back to the NAND gate circuit with the 7400 IC that we talked about earlier.

In calculating the pulse reptition rate (PRR), we use the following equation:

$$\text{PRR} = \frac{1.44}{(R1 + 2R2)C} \qquad (5\text{-}5)$$

Using the values in Figure 4-6, this gives

$$PRR = \frac{1.44}{(1 \times 10^3 + 200 \times 10^3) \times 0.1 \times 10^{-6}}$$

$$= \frac{1.44}{201 \times 10^3 \times 0.1 \times 10^{-6}}$$

$$= 71.6 \text{ Hz}$$

This timer clock circuit is really an astable multivibrator, since it has two momentarily stable states, between which it continuously alternates.

> *Question:*   What do we do if we want a PRR higher than 1 MHz? [We can use an op-amp clock circuit. Some op-amps can operate as high as 100 MHz at full power gain. See the next section.]

OP-AMP CLOCK

We saw earlier how an op-amp could be used as a comparator. In that circuit (Figure 5-4), it was used without feedback. But in the circuit in Figure 5-7 there is feedback to both inputs. A portion of the output voltage appears on the noninverting input, and at the same time the capacitor connected to the inverting input charges or discharges through $R2$.

When V_o is high $V_{i,+}$ is high also, and C charges through $R2$ until $V_{i,-}$ exceeds that level. V_o then goes low. This makes $V_{i,+}$ low also, and C proceeds to discharge via $R2$. When $V_{i,-}$ falls below the level of $V_{i,+}$ V_o goes high, and the cycle repeats.

Figure 5-8 shows $V_{i,+}$ and $V_{i,-}$ superimposed, as viewed on a dual-trace oscil-

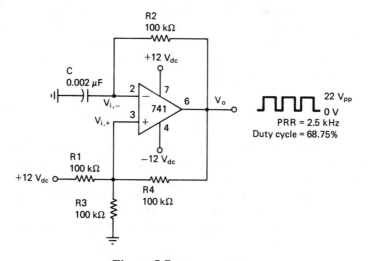

Figure 5-7   Op-amp clock.

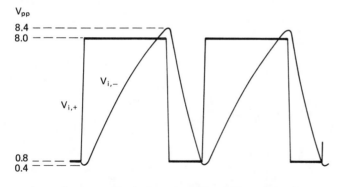

Figure 5-8 $V_{i,+}$ and $V_{i,-}$ waveforms as displayed by a dual-trace scope.

loscope. Each time the value of $V_{i,-}$ rises above or falls below that of $V_{i,+}$, the op-amp output goes low or high, respectively. This gives us a very good idea of the way the op-amp generates a pulse train by being connected as a comparator with feedback.

The PRR is given by

$$PRR = \frac{1}{2(R2)C} \tag{5-6}$$

Using the values in Figure 5-7,

$$PRR = \frac{1}{2 \times 100 \times 10^3 \times 0.002 \times 10^{-6}}$$

$$= 2.5 \text{ kHz}$$

Different values of $R2$ and C will give other values for PRR. These may be calculated by rearranging Equation (5-6).

USING THE 555 IC AS A VOLTAGE-CONTROLLED OSCILLATOR

Since the 555 IC has a terminal (5) for applying a control voltage, it is possible to vary the PRR by changing the voltage, rather than by changing the components responsible for the pulse repetition rate. By connecting a potentiometer and resistor to pin 5, as shown in Figure 5-9(a), we can set various reference voltages on the two comparators. In this experiment, by adjusting the potentiometer we should be able to set the voltage on pin 5 to any value between 1.0 and 5.0 V_{dc}, with consequent PRR values between approximately 350 and 525 kHz. The exact results for different experimenters would depend on the real values of the components used and the accuracy of the PRR measurement.

TROUBLESHOOTING CLOCK CIRCUITS

Troubleshooting clock circuits is made easy if you build a replica on your breadboard. This shows what the circuit should do, and when you compare it with the one with a problem, you can quickly pinpoint the cause.

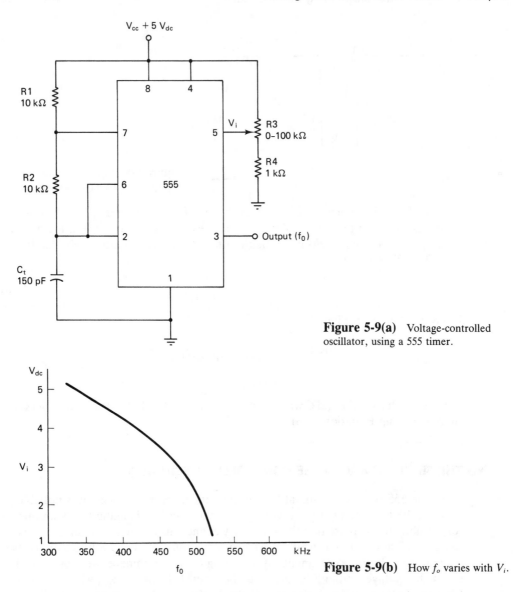

Figure 5-9(a) Voltage-controlled oscillator, using a 555 timer.

Figure 5-9(b) How f_o varies with V_i.

For instance, an astable multivibrator using a 555 IC in a circuit similar to that in Figure 5-6 would normally be delivering a train of pulses at pin 3. However, suppose it does not, and that pin 3 is stuck high with a permanent voltage of 4.5 V_{dc}. Table 5-3 shows how your logic probe would indicate the state of each pin of the 555 IC for various possibilities (shorted resistors are not considered).

Question: What can we do if we don't have a logic probe? [We can use a multimeter, connecting the black lead to ground. Readings above 2 V_{dc}

TABLE 5-3 LOGIC STATES OF THE PINS OF THE 555 IC USING A GLOBAL SPECIALTIES
LP-1 LOGIC PROBE

Pin	Normal	Capacitor Open	Capacitor Shorted	Resistor Open
1	L	L	L	L
2	H	P > 100 kHz	L	Open circuit
3	P	P (duty cycle over 85%)	H	H
4	H	H	H	H
5	H	H	H	H
6	H	P > 100 kHz	L	Open circuit
7	H	P > 100 kHz	Open circuit	H, open circuit[a]
8	H	H	H	H

H, logic state high; L, logic state low; P, pulse.

[a] If $R1$ is open.

are high; those below 0.8 V_{dc} are low. This is not as convenient as a logic
probe, and the multimeter gives no indication of pulse trains.]

The symptom (pin 3 stuck high) would give rise to a column L   H   H   H   H
H   H   H, but this does not match any of the columns in the table. Therefore, we
must assume that the IC itself is to blame.

For another example, we'll use the circuit in Figure 5-3. You will recall that
this circuit puts out a symmetrical pulse. Table 5-4 shows the normal logic states of

TABLE 5-4 LOGIC STATES OF THE PINS OF THE 7400 IC USING A GLOBAL SPECIALTIES
LP-1 LOGIC PROBE

Pin	Normal	C1 Open	C1 Shorted	C2 Open	C2 Shorted	R1 Open	R2 Open
1	L	0	L	L	H	0	L
2	L	0	L	L	H	0	L
3	P^1	0	H	0	L	L	H
4	L	L	H	L	L	L	0
5	L	L	H	L	L	L	0
6	P^1	H	L	P^2	H	H	L
7	L	L	L	L	L	L	L
8	H	0	H	0	H	H	H
9	P^1	H	L	P^2	H	H	L
10	P^1	0	H	0	L	L	H
11	L	0	L	0	L	L	L
12	H	0	H	0	H	H	H
13	H	0	H	0	H	H	H
14	H	H	H	H	H	H	H

P^1, pulse indication; may be either <100 kHz or >100 kHz. P^2, pulse indication; duty cycle >85%.
H, logic state high. L, logic state low. 0, open circuit (no indication on the logic probe).

the 14 pins of this IC when used in this circuit and the effect of various component failures. In addition to those shown, there is also the possibility that one of the capacitors has changed value enough to distort the duty cycle or even to change frequency. In that case the logic state of pins 3 and 6 would suggest an abnormal duty cycle, which we would want to investigate with an oscilloscope.

Let us assume that our circuit is not giving any pulse output. The logic probe indicates an open circuit on pins 1 and 2. We can duplicate this on the breadboard by pulling $R1$ from the circuit. Therefore, we shall expect to find that $R1$ has opened up or become disconnected. But if after simulating all the malfunctions in Table 5-4 we cannot match the symptoms, we should suspect the IC, as before.

Again we can appreciate the value of using a breadboard for troubleshooting. By duplicating the circuit with the problem, we can see how it *should* function, and by comparing the two with each other, the good and the bad, we are led automatically to the faulty component without having to desolder anything first.

6

Digital Logic

The two principal ways in which we use electrons to do things for us are called analog and digital. Sometimes they are called linear and logical, but the first-named terms are the most widely used. Analog means that the profiles of the electric currents and voltages in the circuits correspond to some other physical occurrence. For instance, an audio signal is an electrical analog of a sound, because its voltage rises and falls in a manner corresponding to the rise and fall of air pressure in the sound wave. We can convey information this way by means of a telephone system, in which the sound of the speaker's voice is copied electrically by a microphone and sent over the wires to the receiver, where an earphone uses the electrical analog to create sound waves that replicate the speaker's voice.

In a digital system, the information is changed into pulses that stand for it, but do not replicate it. One use of this method was explained in Chapter 3, where we saw how decimal numbers entered on the keyboard of a computer or calculator are encoded to become binary numbers for internal manipulation and later decoded to become decimal numbers again. Digital computers also convert letters of the alphabet and other symbols into binary numbers, to be handled in various ways and then turned back into a form that we understand.

We can also digitize analog signals. For instance, in a digital audio system the analog audio is changed to digital audio, and then changed back to analog audio before being applied to the speakers. We shall investigate analog to digital conversion and digital to analog conversion in Chapter 10.

We can easily understand the fundamental differences between an analog circuit and a digital circuit by comparing the two transistor circuits shown in Figure 6-1. The analog circuit is an amplifier. We can apply a signal of any form to its input,

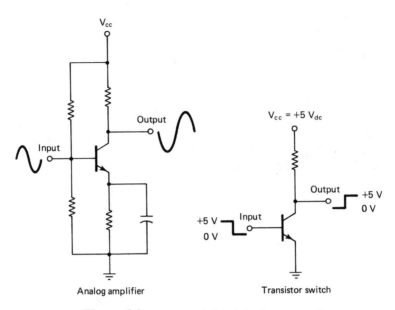

Figure 6-1 Analog and digital circuits compared.

and an enlarged, inverted replica of it will appear at the output. Variations in the
base current cause variations in the bias across the base–emitter junction in the
transistor, resulting in variations in the collector current.

The digital circuit is of a transistor switch. When operated with a V_{cc} of 5 V_{dc},
an input of any voltage between 2 and 5 V will cause the transistor to saturate. In
this condition the resistance offered by the transistor to the collector current is very
low, so the voltage on its collector, that is, its output, is virtually nil.

On the other hand, if the voltage applied to its base is below about 0.7 V, the
transistor's base–emitter junction will not be forward biased, so it is unable to
conduct. Therefore, the collector current is cut off, and no current flows through the
load resistor. Now the full value of V_{cc} appears on the collector.

The transistor switch is activated only by input voltages above 2 V or below
0.7 V.

> *Question:*   How about voltages *between* 0.7 and 2.0 V_{dc}? [The transistor
> switch would respond to these, but we want output voltages of zero or
> 5 V_{dc} only. Therefore, voltages on the base that produce other outputs
> are not allowed. Anything over 2 V_{dc} causes the transistor to saturate,
> giving an output of zero volts.]

The output voltages that are wanted are those that are clearly recognizable as
"high" or "low." These correspond to the 1 and 0 of the binary number system,
which we met in Chapter 2. At that time we considered it only from the point of view
of digitizing decimal numbers and decoding them for seven-segment displays. At this
time we need to go into it a little further.

TRUTH TABLES

Figure 6-2 shows a circuit in which a lamp is controlled by two switches. We can see with no difficulty that both switches must be closed if the lamp is to light. One switch will not do it. All the possibilities of this circuit are explained in Table 6-1, which is called a *function table*.

If we let symbol 0 represent off and symbol 1 represent on, and we also let A, B, and Q stand for the switches and lamp, we can abbreviate this table to that given in Table 6-2, which is called a *truth table*. This is called the AND truth table because A *and* B must both be high for Q to be high.

> ***Question:*** What is the difference between a function table and a truth table? [A function table lists all the possible functions of the components of the circuit, whereas a truth table lists all the possible values of the symbolic logic variables in the circuit. In most logic circuits the variables have only two possible values, 1 and 0. The first is called *true* and the second *false*.]

In the circuit shown in Figure 6-3, Q will be on when either switch A *or* B is closed. It is therefore called an OR circuit, and its truth table is given in Table 6-3.

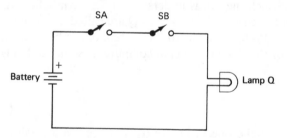

Figure 6-2 Lamp controlled by two switches in series.

TABLE 6-1 FUNCTION TABLE FOR FIGURE 6-2

Switch A	Switch B	Lamp
Off	Off	Off
Off	On	Off
On	Off	Off
On	On	On

TABLE 6-2 THE AND TRUTH TABLE

A	B	Q
0	0	0
0	1	0
1	0	0
1	1	1

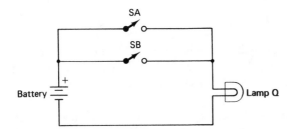

Figure 6-3 Lamp controlled by two switches in parallel.

TABLE 6-3 THE OR TRUTH TABLE

A	B	Q
0	0	0
0	1	1
1	0	1
1	1	1

The symbol · is often used to represent AND, and symbol + to represent OR. In this way we can say $Q = A \cdot B$ or $Q = A + B$. (An alternative form of the first statement is $Q = AB$, in which the dot is understood.) This symbolic notation is called *Boolean algebra*, and the symbols A, B, and Q are called *Boolean (or logic) variables*. George Boole (1815–1864) invented this system of symbolic logic, which today provides the basis for the design of digital computer systems and languages.

GATES

Transistor switches similar to the one in Figure 6-1 are used in a small group of circuits called *gates*. Gates are the building blocks of digital circuits, and practically any digital circuit can be constructed by combining them in various ways. The five most important gates are the inverter, the OR gate, the NOR gate, the AND gate, and the NAND gate.

The inverter is sometimes called a NOT gate, because its output is *not* the same as its input. It is basically the same as the transistor switch in Figure 6-1. In this circuit the output is inverted with regard to the input, and that is what an inverter does. Since a common-emitter amplifier circuit inverts the input, the symbol for an amplifier is used for an inverter, but with a small circle added to emphasize its function as an inverter, as in Figure 6-4.

The Boolean statement for an inverter is $A = \overline{Q}$. The bar over the Q means that it is inverted: when A is high, Q is low, and vice versa. $\overline{Q}$ is called the *complement* or *negation* of Q. If $Q = 1$, then $\overline{Q} = 0$. We shall come to see more uses of this bar later. For the present, we should note that an inverter has one input and one output.

The OR gate's symbol appears in Figure 6-5, with two inputs and one output. Actually, OR gates can have any number of inputs (except one), but only one output. This gate works like the two parallel switches in Figure 6-3: when either, or both, are high, the output is high also. The Boolean statement for this is $A + B = Q$.

If an inverter is added to an OR gate, its output will be the opposite of the OR gate. The OR gate symbol now has a small circle added to indicate the inverter, which is built in. The gate, now called a NOR gate, is shown in Figure 6-6. Its Boolean statement is $A + B = \overline{Q}$. Like the OR gate, its inputs are not limited to two.

The AND gate symbol is shown in Figure 6-7. It works the same way as the two switches in Figure 6-2. Both inputs must be high before the output goes high. The Boolean statement is $AB = Q$. The AND gate is also not limited to two inputs.

The NAND gate is the same as an AND gate with an inverter added. Its symbol is shown in Figure 6-8, and its Boolean statement is $AB = \overline{Q}$. This gate is a very important gate, used more than any other. It is easier to realize in MOS technology than an AND gate. By using NAND gates in the combinations shown in Figure 6-9, any of the previous gates can be made. Thus the NAND gate can be a universal gate, which simplifies and lowers the cost of many digital systems.

Question: Which of these gates should we consider universal? [The NAND gate.]

DIGITAL LOGIC FAMILIES

In their ongoing search for faster and smaller devices, manufacturers of ICs have evolved many different types. However, not all the slower ones have been super-

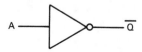

Figure 6-4 Symbol for inverter.

Figure 6-5 Symbol for OR gate.

Figure 6-6 Symbol for NOR gate.

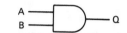

Figure 6-7 Symbol for AND gate.

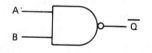

Figure 6-8 Symbol for NAND gate.

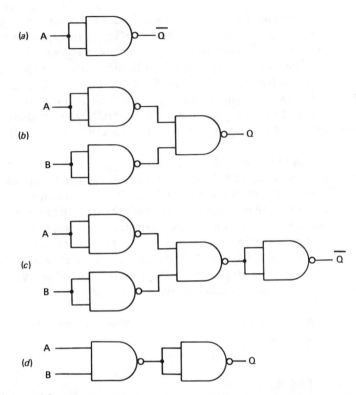

Figure 6-9   How the NAND gate in Figure 6-8 can be used to form any of the other four gates: (a) inverter; (b) OR gate; (c) NOR gate; (d) AND gate.

seded by faster rivals, since some have important advantages that keep them popular. The most well known digital families in use are the following:

Transistor–transistor logic (TTL)
Emitter-coupled logic (ECL)
Integrated injection logic (IIL)
Complementary metal oxide semiconductor (CMOS)

TRANSISTOR–TRANSISTOR LOGIC

Perhaps the best known and most widely used type of logic gate is the bipolar TTL. Figure 6-10 shows the basic TTL NAND gate. The first transistor has more than one emitter and is only turned on when all are high. That makes it an AND gate. The second transistor is an inverter, so the AND gate becomes a NAND gate. As mentioned previously, NAND gates can be combined to form other gates.

TTL circuit voltages are standardized as follows:

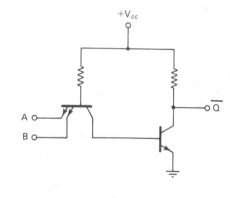

Figure 6-10 TTL NAND gate (the multiemitter transistor can have up to eight emitters).

V_{cc}	5.00 ± 0.25 V
Low (0) input	0 to 0.8 V
Low (0) output	0 to 0.4 V
High (1) input	2.0 to 5 V
High (1) output	2.4 to 5 V

EMITTER-COUPLED LOGIC (ECL)

ECL gates can operate at a much higher frequency (>100 MHz). The frequency limitation for TTL gates arises because they are operated in the saturation mode. The device is either cutoff or saturated. This has a disadvantage in that overdriving the transistors introduces a time delay that does not exist if they are operated in the nonsaturated mode. However, to operate them in the nonsaturated mode would involve biasing the transistors in the linear region, so they would be susceptible to noise.

The basic ECL gate is shown in Figure 6-11. It is a differential amplifier that performs an OR operation on the inputs. The output is amplified by emitter-following transistors, so both the true and complement signals can be made available as outputs with no added delay.

INTEGRATED INJECTION LOGIC

One advantage of unipolar devices such as the CMOS is that they are smaller than bipolar devices (TTL and ECL). Since we want to get as many gates on a chip as possible, this commends them for large-scale integration (LSI). However, bipolar devices generally operate at higher speeds. Figure 6-12 shows a bipolar gate that has the advantage of smaller size, while obtaining the greater speed of a bipolar device. This is an IIL gate (often abbreviated I^2L).

Transistors $Q1$ and $Q2$ are current sources that inject current into the output transistors $Q3$ and $Q4$. When input A goes high, $Q3$ turns on, so its collector goes

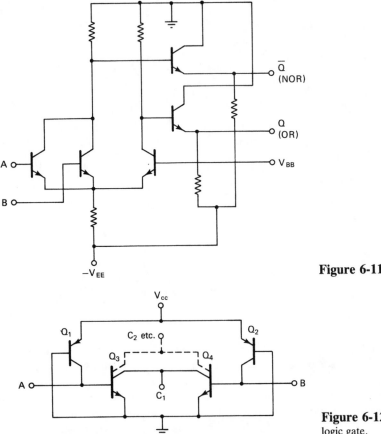

Figure 6-11 ECL OR/NOR gate.

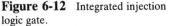

Figure 6-12 Integrated injection logic gate.

low. The same applies to $Q4$ when input B goes high. Since both collectors go to the output $C1$, the output is low whenever A or B, or both, are high. When both inputs are low, $Q3$ and $Q4$ are cut off, so $C1$ is high. This is the NOR function, of course, so this is a NOR gate.

The output voltages are very low. High is 0.7 V (V_{BE}) and low is 0.1 V ($V_{CE,\,sat}$), so an amplifier is usually required for this gate to interface with other logic circuits.

Although the transistors have been portrayed separately, in IC fabrication the source and output transistors are merged. Since there are also no other components, a very high packing density is achieved. Another feature of this gate is that the output transistors may be fabricated with more than one collector to give a multioutput device ($C2$, etc.).

COMPLEMENTARY METAL OXIDE SEMICONDUCTOR

For most LSI, unipolar transistors are preferred to bipolar transistors because they are capable of high circuit density on a chip. They also use less power. Figure 6-13 shows a two-input CMOS NAND gate. $Q1$ and $Q2$ are p-channel enhancement

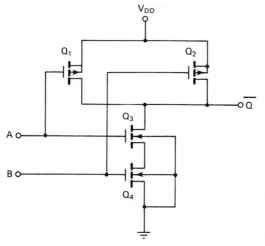

Figure 6-13 CMOS NAND gate.

transistors that are turned on by a low input; $Q3$ and $Q4$ are n-channel enhancement transistors that are turned on by a high input.

If the A and B inputs are both low, then $Q3$ and $Q4$ are both cut off, while $Q1$ and $Q2$ are turned on. The output voltage is then $V_{OUT} = V_{DD}$ = logic 1, or high. If A is low and B is high, $Q2$ is turned off, but $Q1$ is still on; and although $Q4$ turns on, $Q3$ is still off. Therefore, V_{OUT} is still high. But if both A and B are high, $Q1$ and $Q2$ are both turned off, while $Q3$ and $Q4$ are both turned on, so V_{OUT} goes low.

> ***Question:*** What are enhancement transistors? [An enhancement-mode MOS transistor is normally off, but when a gate–source voltage of the appropriate polarity is applied, it drives majority carriers away from the gate area, thus providing a minority-carrier channel between source and drain.]

CMOS circuits are very sensitive to static electricity. It is necessary to take precautions against inadvertently subjecting them to a static charge when handling them. The assembler must be grounded by means of a metal bracelet and ground wire. Soldering irons and other equipment with which they may come in contact must also be grounded. TTL circuits are more rugged and require no special handling.

CMOS circuits operate from a supply of approximately 4 to 15 V. A low (0) is between 0 V and 30% of the supply voltage. A high (1) is between 70% of the supply voltage and the supply voltage. These regions are usually guaranteed for noise immunity, and the higher the supply voltage is, the greater the absolute noise immunity. The region between 30% and 70% is undefined and is not used.

> ***Question:*** If a CMOS IC has a supply voltage (V_{DD}) of 12 V_{dc}, what would be the high and low voltage levels? [A high would be a voltage between 8.4 and 12 V_{dc}. A low would be a voltage below 3.6 V_{dc}.]

The voltage levels in a logic circuit may be either positive or negative. The majority of computers use positive logic, in which logic 1 is chosen as the more positive of the levels. However, some use negative logic, in which logic 1 is chosen as the more negative of the levels.

Either convention can be used, provided it is followed consistently. Since it seems more natural to say that a device (such as a flip-flop) is triggered by a logic 1 rather than a zero, we usually choose the polarity to suit the triggering characteristics of the device.

LOGIC GATE ICs

Figure 6-14 shows the block diagrams and pin arrangements of the most widely used TTL logic-gate ICs:

> 7400 Quad two-input NAND gate
> 7402 Quad two-input NOR gate
> 7404 Hex inverter (six inverters)
> 7408 Quad two-input AND gate
> 7432 Quad two-input OR gate

The 74XX numbering of these devices tells us that they are all TTL devices. An exception would be if the number was in the form 74CXX. This would be a CMOS device, but the latter are usually numbered in a 40XX series, such as 4011, a quad two-input NAND gate. However, some other letters are used with the 74XX series to denote TTL subfamilies, such as the following:

> 74SXX   Schottky device, which is faster
> 74LSXX   Low-power Schottky device, which requires less power than the ordinary 74XX device
> 74FXX   Fast version of the ordinary device

TROUBLESHOOTING TOOLS

Since digital circuits operate by switching between states, from high to low, and vice versa, and since the *exact* value of the potentials is unimportant as long as they are above or below the respective thresholds, the essential equipment we need is a means of verifying the correct voltage *states* on the various pins of an IC or at other circuit points.

Logic Probe

As mentioned before, the most widely used instrument for this is a *logic probe*. The block diagram of a typical one is shown in Figure 6-15. (This is a Global Specialties

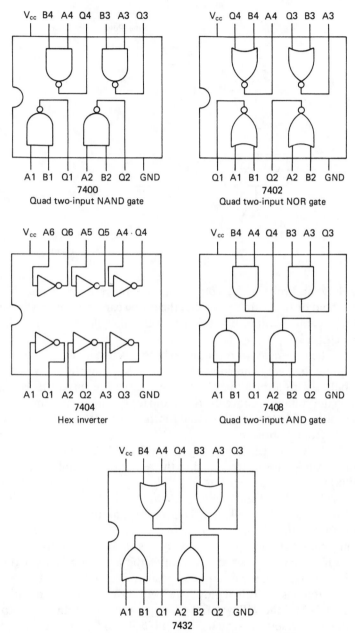

Figure 6-14   The most widely used TTL gate ICs.

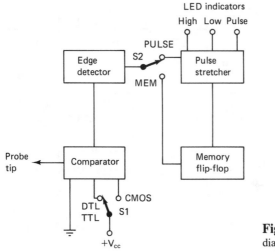
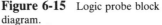

Figure 6-15 Logic probe block diagram.

LP-1 Logic Probe.) The *comparator* reference is set to the TTL or CMOS logic threshold by the switch $S1$ so that the probe can determine whether the voltage state at the IC pin or other point being probed is high or low. Switch $S2$ is normally set to PULSE.

The *edge detector* responds to both positive and negative transitions and drives the *pulse stretcher*. This converts voltage-level transitions, as well as narrower pulses, to pulses with a duration of $\frac{1}{3}$ s. These drive the three LEDs. The red LED lights when the voltage is above the threshold for the high state; the green LED lights when the voltage is below the threshold for the low state; and the yellow LED blinks when a pulse train is present.

The yellow LED blinking by itself indicates a high-frequency (over 100 kHz) pulse train. At lower frequencies, the red or the green LED, or both, may light (not blink) in addition, according to the duty cycle of the pulses. If this is between 15% and 85%, both red and green LEDs will light. But if the duty cycle is less than 15%, only the green LED will light. This means that the signal is normally low and pulsing high for less than 15% of the time of each cycle. If only the red LED lights, the signal is normally high and pulsing low for less than 15% of the time.

If switch $S2$ is set to MEM after the probe tip is applied to the pin or other point being tested, the next event (positive or negative level transition, or pulse, even if the latter has a width of only 50 ns) will activate the memory flip-flop. The flip-flop will latch, and the yellow LED will light and stay lit until the flip-flop is reset. This is done by resetting switch $S2$ to PULSE and then returning it to MEM, while keeping the probe tip applied to the point being tested. If we were to apply the probe tip to this point *after* setting $S2$ to MEM, the flip-flop would be activated when contact was made, giving us a false reading.

Some logic probes have additional features, such as a tone output to save having to shift the eyes from the circuit being tested to the probe LEDs. Others can detect pulses as short as 3 ns and pulse trains as fast as 150 MHz. But generally, they all operate in much the same way, as described above.

Logic Pulser

The logic pulser is a convenient way to inject a single pulse or a pulse train into a circuit. We can use this to stimulate a device that we are testing with a logic probe. These two instruments work together very well for troubleshooting digital circuits.

Logic Clip

The logic clip is constructed along the same lines as a clothes pin. In using one, we open it in the same way, by squeezing the two parts together, so that the smaller, gripping end can be clamped over an 8-, 14-, or 16-pin dual-in-line package IC. The teeth in the jaws of the logic clip make separate contact with each pin, and LEDs on the opposite end light to indicate those in a high state. In this way we can see the state of all the pins of the IC simultaneously.

TROUBLESHOOTING DIGITAL LOGIC

Assuming we have a logic probe, we would begin by probing each pin of the suspected IC and comparing its state with what its truth table says it should be. If we find a discrepancy, we have to try and deduce what might be its cause.

> *Question:* What if we don't have a logic probe? [We can ascertain the dc state of each pin with a multimeter or an oscilloscope, although they are slower, and less convenient.]

IC failures fall into two main categories: internal and external. The most common internal defect is an open bond between the pin and the chip. When this happens to an input bond, the correct signal will be present on the pin, but cannot reach the chip. The input of the chip will float to a "bad" level between the high and low thresholds, which a TTL gate will see as a permanent high state. The effect on the output will depend on the nature of the circuit. If we take as an example a NAND gate with two inputs with a normal truth table, as in Figure 6-16(a), the effect of an open bond at *A* will be the truth table of Figure 6-16(b).

> *Question:* What would happen if the internal bond between pin 7 of a 7448 IC (the decoder we examined in Chapter 3) and the chip was open? [We'd find that the outputs were all odd numbers. Regardless of the input on pin 7, the chip would see a permanent 1 there, so the only outputs we could get from the decoder would be 1, 3, 5, 7, and 9, as if the inputs were 0001, 0011, 0101, 0111, and 1001.]

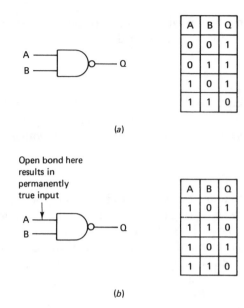

A	B	Q
0	0	1
0	1	1
1	0	1
1	1	0

(a)

Open bond here
results in
permanently
true input

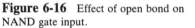

A	B	Q
1	0	1
1	1	0
1	0	1
1	1	0

Figure 6-16 Effect of open bond on NAND gate input.

(b)

An open output bond will block the correct output from the chip so that it will not appear on the output pin. This will also go to a bad level and will affect all input pins on other ICs to which it may be connected. These will behave as if they had open input leads, with the difference that the "bad" level will now be present on all the input pins. This would lead you to look further back for the origin of the malfunction.

Instead of opening up, the input or output bond may short to V_{cc} or ground. If the short is to V_{cc}, all signal lines connected to that point will be high. Conversely, a short to ground will hold them to a low state. The most likely effect of either condition is to inhibit all signals normally to be found beyond this point, so it is one of the easiest troubles to track down.

It is not so easy to analyze a problem caused by an internal short between two pins when neither is V_{cc} or ground. When either pin goes to a low state, it pulls the other with it, yet when the two pins should be high or low together, they show the proper voltage.

An internal failure in the chip is always catastrophic to its performance so that the output pins are locked high or low and will not change in response to appropriate stimuli. This is because the failure blocks the signal flow by completely preventing switching action.

An open signal path in the external circuit has the same effect on the input to which it is connected as an open output bond in the output of the previous IC. The input will float to a "bad" level. However, as the correct signal appears on the output pin, we know that the interruption must exist between the output pin of the preceding IC and the input of the following IC.

A short between the external signal path and V_{cc} or ground exhibits the same

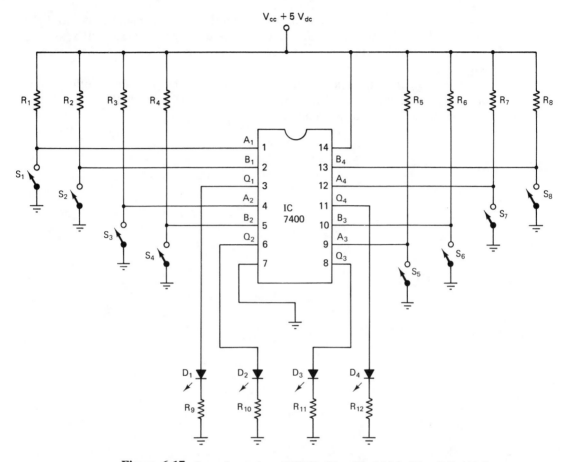

Figure 6-17   Setup for testing a 7400 IC: $R1$ to $R8$, 4.7 kΩ; $R9$ to $R12$, 330 Ω; $S1$ to $S8$, eight-pole DIP switch; $D1$ to $D4$, LEDs. When any switch is open, +5 V$_{dc}$ is applied to the input pin. When any switch is closed, the input is grounded. This ensures that no pin floats to a "bad" input voltage.

symptoms as an internal short of the same kind. If the short is to V_{cc}, the signal path will be high at all times; if to ground, it will be permanently low. This one is hard to isolate, and only a very close inspection of the circuit will determine whether the fault is external or internal.

We can mount a good logic gate IC on our breadboard, apply the various inputs, and see the outputs with an arrangement like that in Figure 6-17. In this example we are examining a 7400 IC. Pins 14 and 7 are used for the common V_{cc} supply and ground, respectively, but the NAND gates are completely independent.

By operating the eight-pole DIP switch in Figure 6-17, we can apply all the high and low inputs in the truth table to each gate and verify that it responds correctly.

The LEDs connected to the outputs light when the output is high and do not light when it is low. The same arrangement can be used for other logic gate ICs, as long as we connect the input voltages to the correct input pins and the LEDs to the correct output pins.

By comparing the operation of our good IC pin by pin with one that is not operating properly, we can deduce what the problem is, and whether it is internal or external, as explained previously.

7

Flip-flops and Counters

The output of a logic gate is immediate and is always the same for each combination of inputs. Logic gates are therefore called *combinational* circuits.

On the other hand, flip-flops can store information, so their outputs do not depend on present inputs only, but also on a sequence of inputs received in the past. For this reason, we call them *sequential* circuits.

But although combinational circuits have no memory in themselves, they are used as building blocks for sequential units, which do have memory. An example of this is shown in Figure 7-1. Here five logic gates (an inverter, two AND gates, and two NOR gates) are connected to form a *gated latch*. This circuit is widely known as an *R–S latch*.

R–S LATCH

Information appearing at the D input is passed straight through to the true output Q whenever the gate input G is high. The second output $\overline{Q}$ is always the opposite, or complement, of Q and has a bar over it to distinguish it. We call it "not-Q" or "Q-bar."

When input G is changed to 0, however, the output is *latched*, or frozen, and will not change in spite of changes on input D. Thus, one bit of data is stored for as long as we keep G low.

The two ANDs and the inverter on the left are called the *gating network*. When G and D are high, the output of the lower AND is 1, and the output of the upper

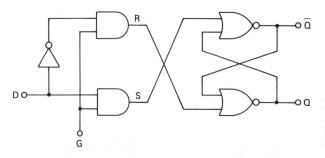

Figure 7-1 *R–S* latch (gated latch): *D*, data input; *G*, gate input; *R*, reset line; *S*, set line; *Q*, true output; $\overline{Q}$, complement output.

AND is 0. In other words, the *S* line is high and the *R* line is low. However, when *G* is low, the outputs of both ANDs are low.

> *Question:* In most flip-flops the gating signal is called a clock signal. Why is this? [Because it keeps the timing of the changing of the data in various parts of the system in step.]

When *S* is high the inputs to the upper NOR gate are both 1, since *Q* is also high. Its output is therefore low. At the same time the inputs to the lower NOR gate are 0 and 0, since $\overline{Q}$ is low and *R* is low. Its output is therefore 1.

But when *G* is 0, both *S* and *R* are 0. The inputs to the lower NOR gate are also both 0, so its output is 1. The inputs to the upper NOR gate are 1 and 0, so its output is 0. Nothing has changed, and nothing will change as long as *G* is 0. Hence a 1 bit is stored in the flip-flop until it is unlatched.

This NOR flip-flop gets its name *R–S latch* from the names of its two inputs "reset" and "set." They are so named because, if *R* is 0, a 1 on *S* will *set Q* to 1; and if *S* is 0, a 1 on the *R* line will *reset Q* to 0.

We call the entire circuit in Figure 7-1 a *gated latch* for obvious reasons. It is a simple, economical unit, but it has one drawback. When *S* and *R* are both 1 before a clock pulse, *Q* may be *either* 1 or 0 after the clock pulse. As long as we don't allow this to happen, the simple *R–S* latch has many uses (such as in some shift registers).

> *Question:* What keeps the outputs of an *R–S* latch from changing when one input changes to 0 while the other remains at 0? [The output of each NOR gate is fed back to another input on the other gate.]

We shall discuss shift registers later, but there is one important fact to consider about them right now. Data must be transferred from one flip-flop to another with both triggered by the same timing signal. To avoid confusion, we need flip-flops that can store their input signal before they change their output signal. We can best do this by using a *pair* of latches instead of a single one. Such an arrangement is known as a *master–slave* flip-flop.

MASTER–SLAVE FLIP-FLOP

Figure 7-2 shows two *R–S* latches connected so that one drives the other. The driver is called the *master*. The driven one is called the *slave*. The gating signal is now called the clock signal, because it is the timing signal fed to both latches.

We are also using three-input NOR gates in the master flip-flop. The extra inputs are called *preset* and *clear*. They are normally kept at 0 and then have no effect on the operation of the flip-flop. Not all flip-flops have these inputs, but they may be used to clear all old data from a register, so that only zeros are stored, or preset it to all ones.

We have also removed the inverter from the gating networks, so we have two data inputs. A new inverter is installed to invert the clock pulse applied to the slave flip-flop.

In this arrangement, when a clock pulse is received by the master, it accepts a new bit of data, but the same clock pulse, because it is inverted, prevents the slave from releasing the bit it is holding. Then, when the clock goes back to zero, the master holds the new bit and the slave releases the old one and receives the new one in its place.

Question: What is the purpose of the master–slave feature in some flip-flops? [To let the flip-flop accept new data before changing the output.]

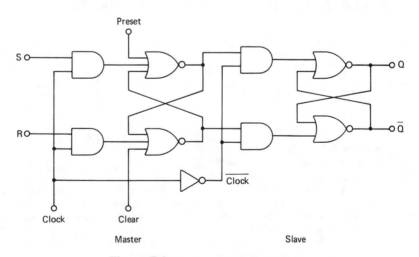

Figure 7-2 *R–S* master–slave flip-flop.

D FLIP-FLOP

Figure 7-3 shows an *R–S* master–slave flip-flop with an inverter between the *S* and *R* inputs. This ensures that the *R* input cannot be 1 at the same time as the *S* input. The *S* input is now called the *D* (*data*) input.

As long as a clock pulse is present on the *C* input, *Q* cannot change, regardless of what is on the *D* input. But when the clock input goes from 1 to 0, the *Q* output becomes the same as whatever is on the *D* input.

> *Question:* Which kind of flip-flop has only one data input? [The *D* flip-flop.]

T FLIP-FLOP

The *T* stands for *toggle*. In Figure 7-4 we see that an *R–S* flip-flop has been altered so that it has no data inputs at all. Its outputs change state on each clock pulse. This is called *toggling*. Some *T* flip-flops toggle when the clock pulse goes from 0 to 1 and others when it goes from 1 to 0, according to their design.

J–K FLIP-FLOP

In Figure 7-5 we have altered the master–slave *R–S* flip-flop of Figure 7-2 so that the input AND gates now have three inputs each. The extra inputs on the AND gates are cross-connected to the *Q* and $\overline{Q}$ outputs. If both the *J* and *K* inputs (formerly *S* and *R*) are high, the flip-flop toggles like a *T* flip-flop.

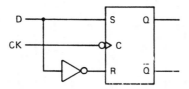

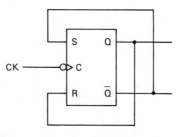

Figure 7-3 *D* flip-flop. It consists of an *R–S* master–slave flip-flop (see Figure 7-2) with an inverter connected between the *S* and *R* inputs. The *S* input thus becomes the data (*D*) input, and it is impossible for both inputs to be high or low at the same time. The triangle on the clock input (CK) denotes that the circuit responds only to a *change*; the circle denotes that the change must be from high to low.

Figure 7-4 *T* flip-flop. The *Q* and $\overline{Q}$ outputs are cross-connected to the *S* and *R* inputs. The outputs change state when the clock input goes from high to low in this flip-flop. This is called toggling.

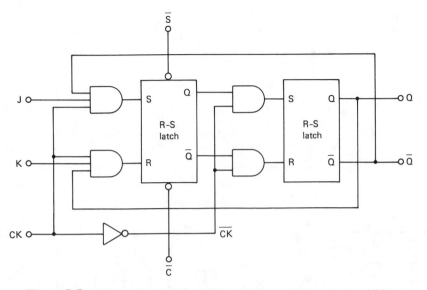

Figure 7-5 *J–K* flip-flop. This is an *R–S* master–slave flip-flop, with the addition of feedback like the *T* flip-flop, but retaining the original two inputs (now renamed *J* and *K*).

If the *J* and *K* inputs are not held high, the flip-flop behaves like a clocked *R–S* flip-flop, but the feedback connections eliminate the uncertain condition that arises in an *R–S* flip-flop when both inputs are high.

Question: When both the *J* and *K* inputs are held high, what kind of flip-flop does the *J–K* flip-flop behave like? [The *T* flip-flop.]

There are various versions of the four basic types of flip-flop we have been examining. Most of the variations pertain to what happens when the clock signal goes from 0 to 1, and 1 to 0, and so on. The actual sequence for any flip-flop is spelled out in the truth table given in the manufacturer's specification for the product.

ASYNCHRONOUS BINARY RIPPLE COUNTER

A counter is made from a series of flip-flops. In Figure 7-6, a ripple counter is shown consisting of four *J–K* flip-flops. Notice that all the *J–K* inputs are held high. The $\overline{\text{CLR}}$ input is first made low so that all the flip-flops are cleared and their *Q* outputs set to 0. It is then held high so that counting may begin.

Nothing happens when the first pulse to be counted arrives at the clock input as long as C_0 is high, but when it reverts to zero (on the 1 to 0 transition of the pulse), the flip-flop toggles, and its Q_0 output becomes 1. This 1 is also applied to the C_1 input of the next flip-flop, but nothing happens to that flip-flop as long as its clock

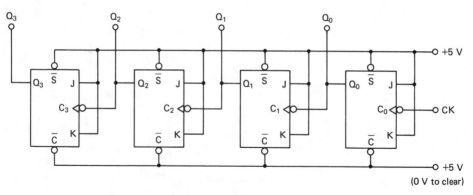

(a) Asychronous ripple counter

P	Q_3	Q_2	Q_1	Q_0
0	0	0	0	0
1	0	0	0	1
2	0	0	1	0
3	0	0	1	1
4	0	1	0	0
5	0	1	0	1
6	0	1	1	0
7	0	1	1	1
8	1	0	0	0
9	1	0	0	1
10	1	0	1	0
11	1	0	1	1
12	1	1	0	0
13	1	1	0	1
14	1	1	1	0
15	1	1	1	1

(b) Q outputs

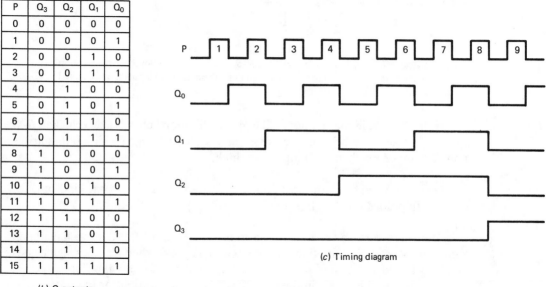

(c) Timing diagram

Figure 7-6   Ripple counter made with four *J–K* flip-flops.

input remains high. The outputs of the four flip-flops are as shown in the second line of the table in Figure 7-6(b), which is binary 0001. In other words, the counter has now counted *one* pulse and is storing the result.

When a second pulse causes the first flip-flop to toggle again, Q_0 goes to 0, so C_1 becomes zero also. The second flip-flop toggles, and its Q_1 changes to 1. This 1 also appears on the C_2 input of the third flip-flop. The outputs of the four flip-flops are now as shown in the third line of the table in Figure 7-6(b), or binary 0010. In other words, the counter has now counted *two* pulses and is storing the result.

When a third pulse causes the first flip-flop to toggle again, its Q_0 output goes to 1, and so does the C_1 input of the second flip-flop, but the Q_1 output of the latter stays at 1 as long as the input is high. The outputs of the four flip-flops are now as shown in the fourth line of the table, or binary 0011. In other words, the counter has now counted *three* pulses and is storing the result.

When a fourth pulse causes the first flip-flop to toggle again, its Q_0 output goes to 0, and so does the C_1 input to the second flip-flop. This toggles in turn, and its Q_1 output goes to 0. This causes the third flip-flop to toggle, and its Q_2 output goes to 1. This 1 also appears on the C_3 input of the fourth flip-flop. The outputs of the four flip-flops are now as shown in the fifth line of the table, or binary 0100. In other words, the counter has now counted *four* pulses and is storing the result.

Succeeding pulses cause the four flip-flops to toggle as shown in the table, and with each input pulse the count increases by 1 until all four Q outputs are 1, equivalent to decimal 15. On the 1 to 0 transition of the next pulse, *all* flip-flops toggle, since all inputs go to 0, and the counter output is now 0000.

This counter, therefore, counts from 0 to 15, and resets to 0. If we wanted it to count higher, we could add more flip-flops. It gets the name *ripple counter* because the changes caused by the input pulses ripple through the flip-flops from right to left. As each flip-flop takes a certain amount of time to respond, it is not possible for them all to respond simultaneously. The actual time by which each lags the first depends on its position in the series. Because of this, the ripple counter is an *asynchronous* counter.

It is also a *divide-by-16* counter, because the output of the fourth flip-flop is only 1 pulse for 16 input pulses. It can therefore be used as a frequency divider, as shown in the timing diagram.

> *Question:* What are the features of an asynchronous binary counter? [It is made of T flip-flops; it changes its outputs in ripple fashion and is useful as a frequency divider.]

SYNCHRONOUS COUNTER

A *synchronous* counter (sometimes called a *parallel* counter) is shown in Figure 7-7. In this counter the clock inputs of the $J–K$ flip-flops are connected together so that the pulses to be counted arrive at all of them simultaneously.

As we have seen before, this $J–K$ master–slave flip-flop toggles on the 1 to 0 transition of each pulse to be counted as long as both J and K inputs are high. We must, therefore, arrange the circuitry of this counter so that the inputs of each flip-flop are in the proper state to allow it to toggle when we want it to, but not when we don't.

To achieve a binary count, the Q outputs of the flip-flops must go high and low in accordance with the timing diagram in Figure 7-7. The first flip-flop has to toggle on the 1 to 0 transition of each pulse, so its J and K inputs are held high permanently. The second must toggle on the 1 to 0 transition of alternate pulses, so its J and K

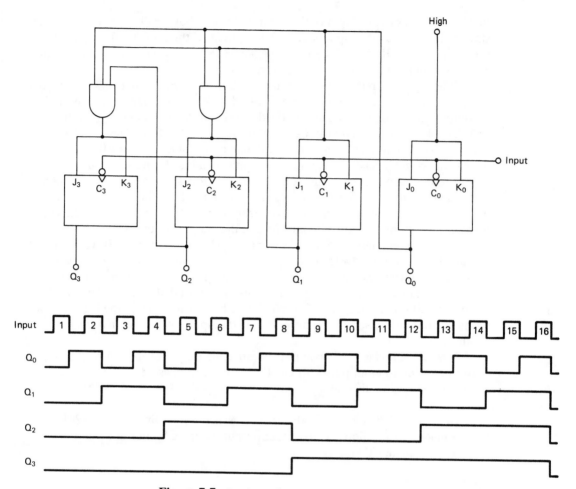

Figure 7-7 Synchronous counter and timing diagram.

inputs are connected to the Q output of the first flip-flop. In this way, it can respond only to every other pulse to be counted.

The third flip-flop has to respond to the 1 to 0 transition of every fourth pulse to be counted. This is achieved by using an AND gate. The AND gate's output goes high when both its inputs are high, which happens when both the Q outputs of the first and second flip-flops are high. Then they go low, so the flip-flop cannot toggle until they are high again, which is after another four pulses, as shown in the timing diagram.

For the J and K inputs of the fourth flip-flop to be high, all the Q outputs of the first three flip-flop's must be high. This doesn't occur until the eighth pulse to be counted arrives, after which they all go low, and are not all high again for another eight pulses. Consequently, the Q output of the fourth flip-flop stays high until that happens.

When the sixteenth pulse to be counted arrives, the Q outputs of all four flip-flops go low, and the counter recycles.

Question: How do we make a counter synchronous? [By "steering" each flip-flop to the correct next state by decoding the state before.]

Question: What are the advantages and disadvantages of asynchronous and synchronous counters? [In an asynchronous counter, each stage is clocked by its preceding stage, so that the command signal ripples through the chain of flip-flops. This causes a delay, so asynchronous counters are relatively slow. However, as explained above, the counter is useful for frequency division. In a synchronous counter, all the flip-flops are clocked simultaneously, which gives a faster counter, although it is more complex.]

DECADE COUNTERS

The two counters we have been discussing counted from 0 to 15 and then recycled. However, if we want a counter to give a decimal readout, it should count only from 0 to 9. Figure 7-8 shows a ripple counter with the addition of an AND gate to bring this about.

In this circuit, all the J and K inputs are held high except for the J input of the fourth flip-flop. The first three flip-flops, therefore, count up to 7 in the same way as in the asynchronous counter of Figure 7-6. At this point, the Q outputs of all three flip-flops are high, so both inputs of the AND gate are high, placing a 1 on the J input of the fourth flip-flop. This flip-flop is now able to toggle when the Q output of the first flip-flop (which is applied to its clock input) goes low. The Q outputs of the second and third flip-flops also go low, causing the J input of the fourth flip-flop to go low also; so only the Q output of the fourth flip-flop is high.

The $\overline{Q}$ output of this flip-flop is connected to the J input of the second flip-flop, and because it is the complement of the Q output, it is low. The second flip-flop is thereby prevented from toggling when the next pulse makes the Q output of the first flip-flop high. We thus have a count of 9. When the Q output of the first flip-flop goes low, the clock input of the fourth flip-flop goes low also. Since the J input is low, but the K input is high, this makes the Q output low as well; so all flip-flops are now back to 0 and ready to recycle.

This action of the fourth flip-flop takes place on the 1 to 0 transition of every pulse, except when the count of 7 makes its J input high, so that its Q output is low at all other times.

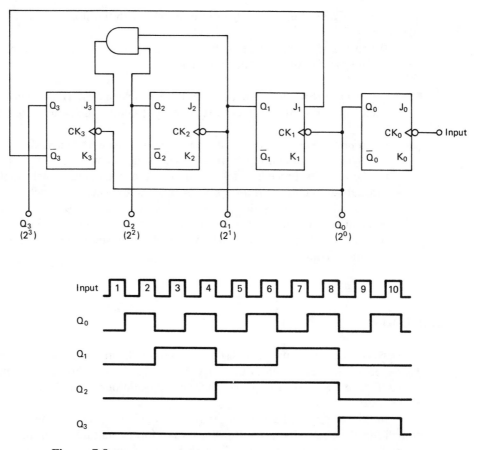

Figure 7-8 Decade counter and timing diagram: J_0, J_2, K_0, K_1, K_2, and K_3 held high; J_1, J_3 connected as shown (J_1 is high as long as Q_3 is high).

INTEGRATED COUNTERS

All three of the counters we have been looking at are made with two 7476 J–K flip-flops. However, in practice we would use counters in which all flip-flops and other elements of the counter circuit were integrated in a single package. Some of these have a binary output, with four pins that can be connected to a decoder to obtain a decimal reading. In others, the decoder is built in so that the output may be connected directly to a seven-segment readout.

The following are examples of integrated counters:

4017	Decade counter/divider
7490	Divide-by-2 or 5, binary-coded decimal counter
7492	Divide-by-12 counter

7493   Binary counter
74160   Decade counter
74161/3   Synchronous presettable binary counters
74190   Up–down decade counter
74191   Up–down binary counter
74192   Up–down decade counter
74193   Up–down binary counter

Except for the first in the list, all these ICs are TTL, operating on +5 V_{dc}. The 4017 is CMOS, for which the maximum supply voltage is 16 V_{dc}.

TROUBLESHOOTING COUNTERS

Suppose we have a ripple counter like that in Figure 7-6, with its Q outputs connected to a 7448 IC decoder, which in turn drives a seven-segment display, as shown in Figure 7-9(a). There is something wrong with this circuit, since the display cycles from 8 to 15 only. When we duplicate the circuit on our breadboard (Figure 7.9(b)), it works perfectly, so we must look for a discrepancy between the two circuits.

A preliminary check with a logic probe shows us that the Q_0, Q_1, and Q_2 outputs of the counter are normal, but that the Q_3 output is either an open circuit or an out-of-tolerance signal, since the probe LEDs do not light at all when the probe is applied to this pin (pin 11 of the second 7476). Measuring voltages with a multimeter gives the results shown in Table 7-1.

Question: Since there is only one seven-segment display unit, how does it show numbers 10 through 15? [It doesn't. The display shows:

TABLE 7-1   COUNTER OUTPUT VOLTAGES

Display	Q_3	Q_2	Q_1	Q_0
8	1.5	0	0	0
9	1.5	0	0	3.7
10	1.5	0	3.7	0
11	1.5	0	3.7	3.7
12	1.5	3.5	0	0
13	1.5	3.5	0	3.7
14	1.5	3.5	3.7	0
15	1.5	3.5	3.7	3.7

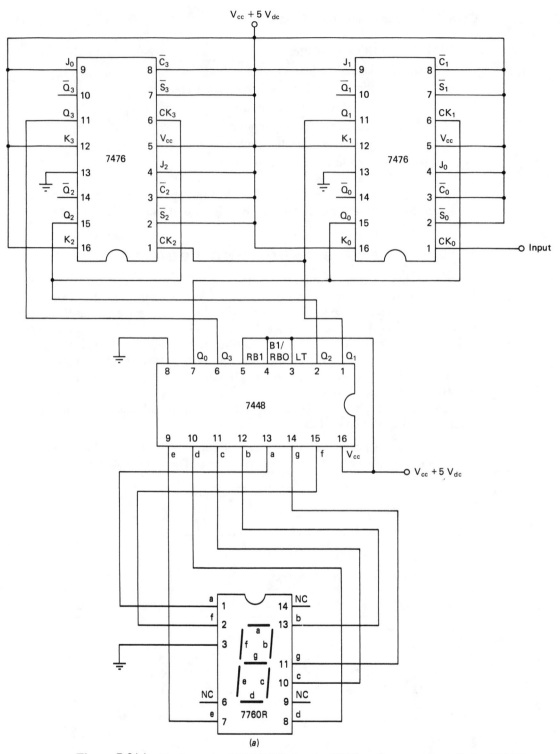

Figure 7-9(a) Ripple counter (7476, 7476), decoder (7448), and seven-segment display (7760R).

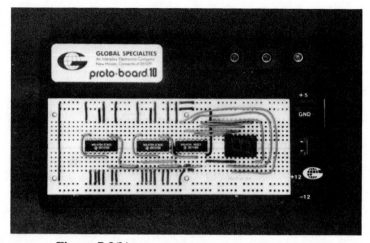

Figure 7-9(b)   Breadboard realization of Figure 7-9(a).

Obviously, the Q_3 output is permanently at a bad level, since it is between 0.4 and 2.4 V (neither low nor high). Nevertheless, this looks like a high to the 7448, so it is decoding the Q outputs accordingly.

We can simulate this on the breadboard by disconnecting the Q_3 output from the 7448 input. The same fault appears, because the 7448 input floats to a bad level if nothing is connected to it. However, if we ground pin 6 of the breadboard 7448, the indicated count changes to 0 through 7, which is correct. Doing this in the circuit with the problem produces the same effect.

The deduction, therefore, is that there is an open bond between pin 11 of the second 7476 and the chip. This is an internal problem, so we must replace the IC. When this has been done, we get a normal count from 0 through 15.

8

Memory

REGISTERS

In Chapter 7 we saw how flip-flops are combined to form counters, and we mentioned that they can also be used in *registers*. Registers are important components of microprocessors, in which they are used as short-term storage devices, each consisting of an array of flip-flops with the capacity to store a computer word. Since each flip-flop stores one data bit, the number of flip-flops in a register depends on the size of the computer word handled by the microprocessor. Registers are either *parallel* or *shift*.

> *Question:* What is a computer word? [The number of bits in a sequence that is handled as a unit and that normally can be stored in one location in a memory.]

A parallel register is shown in Figure 8-1. The set of switches S_0 through S_3 can be set to apply any 4-bit data word to the register, which consists of four J–K flip-flops. The $\overline{CLR}$ line is grounded momentarily to clear the register, and then a clock pulse is applied to the CLK line and simultaneously to each flip-flop. When the clock pulse returns to zero, the Q output of each flip-flop goes high or low according to what is on its J input.

We can remove the input signal from the J inputs and apply a different one, but the data stored in the register are not changed. To store other data, we must ground the $\overline{CLR}$ line, and then clock in the new data. The data enter this register

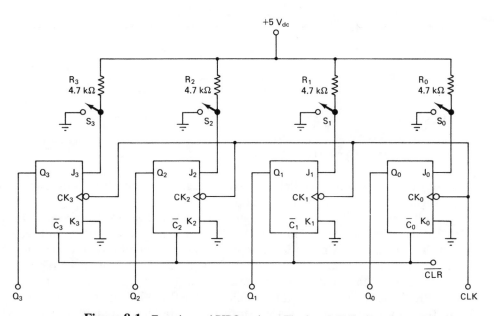

Figure 8-1 Experimental PIPO register. The four J–K flip-flops (two 7476 ICs) are connected as shown. The $\overline{\text{CLR}}$ line is first grounded to clear the register and then held high. (The S pins, not shown, are also held high.) Any 4-bit data word is applied to the J inputs by setting S_0 through S_3. The trailing edge of the next clock pulse causes whatever is on each J input to appear on the corresponding Q output. (The $\overline{Q}$ outputs are not used.) The data word is now stored in the register. Another data word can be stored only after clearing the register by grounding the $\overline{\text{CLR}}$ input.

in parallel and emerge in parallel. It is therefore called a parallel-in, parallel-out (PIPO) register.

A shift register is shown in Figure 8-2. The data enter in series at the J input of the first flip-flop, one bit at a time, on each successive clock pulse. It takes four clock pulses, therefore, to load this register, whereas the parallel register required only one. On the trailing edge of the first pulse, the Q output of the first flip-flop goes high or low, according to what was on its J input. This output now appears on the J input of the second flip-flop, so on the trailing edge of the next clock pulse it causes the Q output of the second flip-flop to change accordingly.

While this is going on, the next data bit on the J input of the first flip-flop is clocked in, and so on, until all four flip-flops are storing data bits. Further clock pulses will move these bits along the register and out through the Q output of the fourth flip-flop. The output will be lost if we do not provide some place for it to go. One thing we can do is to connect the last Q output to the first J input so that the data go round and round.

This kind of register is called a serial-in, serial-out (SISO) register. We can, however, take the data out in parallel, as in the parallel register in Figure 8-1. In that

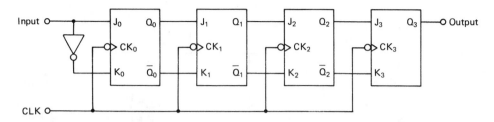

Figure 8-2   Serial-in, serial-out (SISO) shift register. The S and C lines are not shown, but are held high.

case, it is a serial-in, parallel-out (SIPO) register, which has the capability of changing serial data into parallel data.

It is also possible to enter parallel data into a shift register, as shown in Figure 8-3. When the *load-control* input is 0, the circuit functions as a plain shift register, receiving data serially at the first S input. However, when the load-control input is 1, parallel data are entered at the PRESET and CLEAR inputs. Each flip-flop has an AND gate connected to each of these inputs. A 1 data bit causes a 1 output from the PRESET AND gate, and this is applied to the PRESET input; then the Q output of that flip-flop becomes a 1, thus storing a 1 data bit. The input data bit is also inverted and applied to the CLEAR AND gate, which does not go high, so a 0 is applied to the CLEAR input. When a 0 data bit is received, the PRESET AND gate output remains low; but the negative bit is inverted by the inverter, and so the CLEAR AND gate output goes high. This causes the Q output to go low, thus storing the 0 data bit.

This kind of register is called a parallel-in, serial-out (PISO) register. Used in conjunction with a SIPO register, it makes it possible for parallel data to be changed to serial data for transmission along a single wire and for the serial data to be changed back to parallel data later.

LARGER MEMORIES

Memories using flip-flops require a flip-flop for each bit stored. For a memory of any size, this will take up a lot of space and use a lot of current. However, such a memory has one advantage: as long as the power is applied, the flip-flops will continue to store the bits in them. This kind of memory is called a *static RAM*.

RAM means *random-access memory*, which is a storage arrangement into which information can be written or from which it can be read, regardless of where it is located. This differs from a magnetic-tape memory, since with a tape we would have to start at the beginning and run through the tape sequentially until we came to the part we wanted. With a RAM we are free of the sequencing and can go directly to any location, independently of where it is or where we went before.

Most RAMs on chips are *dynamic RAMS*, however, since these take up much less space. They consist of memory cells that store four or more bits each.

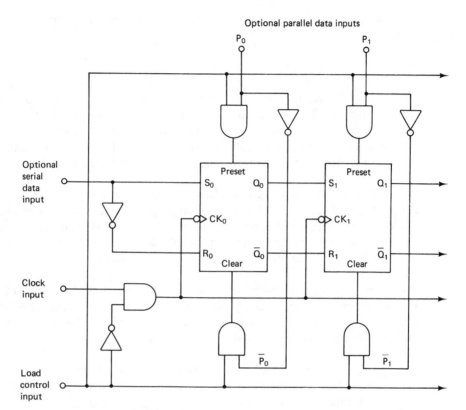

Figure 8-3 Shift register with load control, using *R–S* flip-flops. When load-control input is 0, data at the serial data input are clocked through serially, as in the shift register in Figure 8-2. When load control is 1, clock pulses are blocked, and data are entered at the parallel data inputs. When the load-control input returns to 0, the data are shifted serially. (Only the first and second flip-flops are shown.)

Each bit of data in a dynamic RAM memory cell is stored in a minute capacitor as a charge, or lack of charge. The charge leaks away quickly. The memory chip deals with this problem by *"refreshing"* the charge at frequent intervals. It does this by reading all the memory cells and rewriting their contents before the data can disappear.

Each instruction or data word is written into a specific *address*, where it can be read at a later time when the information is needed. In the microprocessor (control unit), the memory address register (MAR) holds the address of the word currently being accessed, and the memory data register (MDR) holds the information being written into or read out of that location. A *write* command causes the data in the MDR to be stored at the location indicated by the address in the MAR. A *read* command causes the data at the location indicated by the MAR to be transferred to the MDR. The data stored in memory are not destroyed by being read, but any data that were previously in a location into which new data are written are destroyed.

> *Question:* How does the MAR find the RAM cell it wants? [The storage cells are arranged on the chip in a square pattern of rows and columns. Part of the address controls the row decoder, which then activates the desired row. That makes all the cells in that row transmit their bits to their column lines. The other part of the address controls the column selector, which then activates the columns for the desired data word. The data are then transmitted to the word output register and thence to the MDR.]

RAM is used to store variable data. Read-only memory (ROM) is used to store constant data. Data in RAM are lost when power is removed (which is why we call RAM *volatile*), but ROM is permanent. This is because the data are programmed into the chip during fabrication and are in physical or magnetic form. They are not lost (*nonvolatile*) when power is removed. They are accessed like data in RAM (see Table 8-1).

Figures 8-4(a) and 8-4(b) show how a RAM chip would be connected as the internal memory of a microcomputer. The IC here is a 2114 static RAM with 1024 memory cells. Each cell stores 4 bits, and there are 10 address lines, but in this example we are using only 4 (pins 4 through 7).

> *Question:* What do we do with the other address lines that we are not using? [They should be connected to ground; otherwise, these inputs will be misinterpreted by the chip.]

Suppose this memory chip interfaces with a microprocessor, as shown in Figure 8-5. There are three buses.

> *Question:* What is a bus? [A bus is a path through which data, addresses, or control signals can be transmitted and received. When a data bus has arrowheads at each end, it means that data can pass either way, and it is called a *bidirectional* bus. If there is only one arrowhead, it is a *unidirectional* bus.]

In the circuit of Figure 8-4 both the address bus and the data have four lines. The sets of switches on the left simulate the MAR and MDR of a microprocessor. Control signals from the microprocessor are applied to the $\overline{CS}$ and $\overline{WE}$ pins. These would take two lines if the memory were connected to an actual microprocessor, but in this example we shall just apply the required voltage directly to them.

One very important aspect in the memory interface, which is vital for successful operation, is *timing*. This is shown diagrammatically in Figure 8-6. When writing data into memory, the system clock ensures that address and data words are loaded into the MAR and MDR first. Then, when these are stable, the $\overline{WE}$ control signal

TABLE 8-1 SEMICONDUCTOR MEMORY TECHNOLOGY

Type	Construction	Advantages	Disadvantages	Application
		Volatile (Data Lost When Powered Down)		
Dynamic RAM	Single MOS transistor cells	High density, lower cost, lower power	Lower speed; requires refreshing	Main memory
Static RAM	Flip-flop cells (MOS or TTL)	Higher speed; does not require refreshing	Lower density, more power, higher cost	Main memory
		Nonvolatile (Data Not Lost When Powered Down)		
ROM	Mask-programmed by manufacturer	Permanent	Cannot be reprogrammed	Code converter, look-up tables, and so on
PROM	Fusible links	Permanent	Cannot be reprogrammed	Code converter, look-up tables, and so on
EPROM (EROM)	Floating gate cells charged by tunneling[a]	Can be erased by UV light	Must be removed to erase and reprogram	Semipermanent memories
EEPROM (EEROM)	Floating gate cells charged by tunneling[a]	Can be erased electrically without removal from circuit[a]		Semipermanent memories

[a]Tunneling of electrons between floating gates and substrate is brought about by raising the select and program lines to 20 V_{dc} while the columns are grounded. To write data in, the select and column lines are raised to 20 V_{dc}, while the program line is at 20 or 0 V_{dc} for a 0 or 1 bit as required.

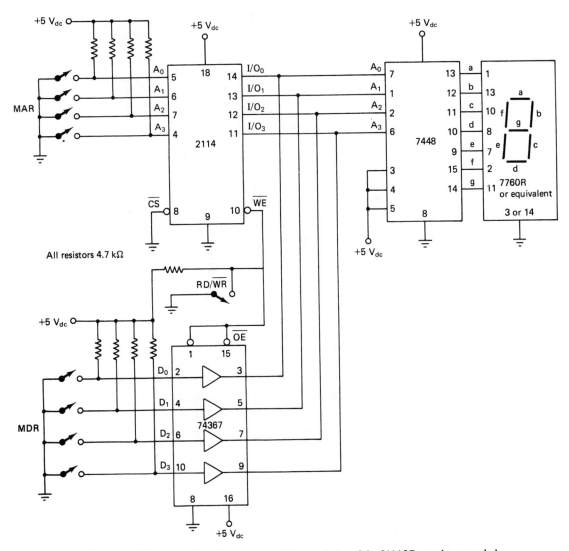

Figure 8-4(a) Static RAM experiment. All unused pins of the 2114 IC *must* be grounded.

(low) is sent to the memory, and the $\overline{CS}$ control signal (low) to enable the memory. The data on the bus lines are now entered into the memory at the addressed location. They overwrite the previous contents of the memory, which are lost.

> *Question:* What would happen if the address lines on the 2114 were changed before the $RD/\overline{WR}$ line was returned to the RD (high) condition? [The same data would be written into two locations.]

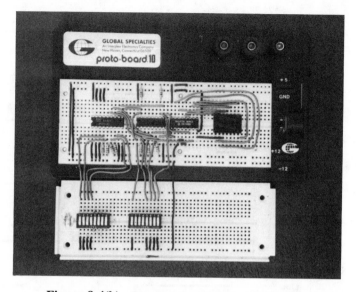

Figure 8-4(b) Breadboard realization of Figure 8-4(a).

To read a location in the memory, the $\overline{WE}$ line is set high. The chip is then enabled by the $\overline{CS}$ line going low. The memory contents at the addressed location are transferred to the output pins and remain there until $\overline{WE}$ or $\overline{CS}$ changes. Reading the memory is *nondestructive*. The contents of the memory cell are not changed by connecting it to the output pins.

On some memory chips the $\overline{WE}$ (write enable) pin is called $\overline{WR}$ (write/read), and the $\overline{CS}$ (chip select) pin is called $\overline{CE}$ (chip enable). If there are several memory chips, the $\overline{CS}$ control signal goes to the one where that address is located.

The circuit in Figure 8-4 contains a noninverting tristate buffer (74367) between the MDR and the RAM (2114). This IC consists of four identical noninverting buffer amplifiers controlled by an $\overline{OE}$ signal (pin 16). When the $\overline{WE}$ control signal goes low (for a write), the $\overline{OE}$ control signal also goes low to enable the buffer. Data in the

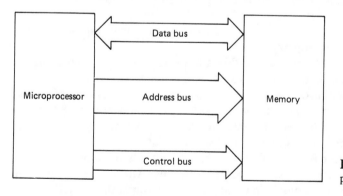

Figure 8-5 Buses between micro-processor and memory.

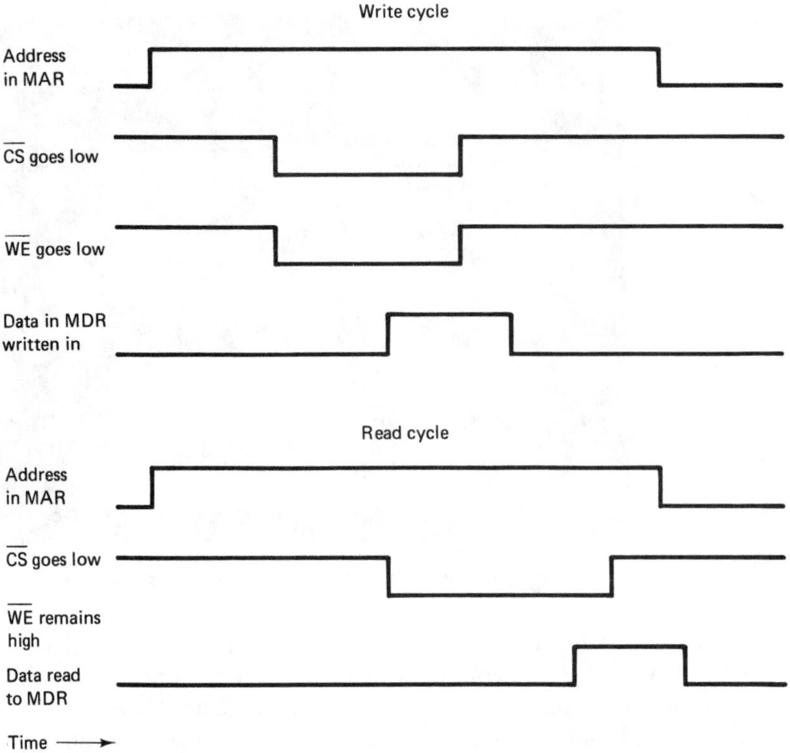

Figure 8-6   Memory timing chart. Timing charts are supplied with memory chips showing the sequence of events, the duration of each, and the intervals between events required for proper operation. The timing is controlled by the system clock.

MDR are then passed through to the input–output data pins (11 to 14) of the memory chip. But when the $\overline{WE}$ control signal goes high (for a read), the buffers are disabled, and their outputs will become a high impedance, so the data read out of the memory do not go back to the MDR. This is why this type of buffer is called a tristate buffer. When enabled by a low $\overline{OE}$ signal, the outputs will be high or low, corresponding to the inputs. But when the $\overline{OE}$ signal is high, the outputs will represent a high impedance regardless of the inputs.

However, in a microprocessor we would want the data output from reading the memory to be able to go to the MDR, so we would not use this type of buffer. Instead, we would use a bidirectional buffer, or *transceiver*. In this type of buffer, there are two sets of buffers. When one is enabled, the data are passed in one direction, but when the other is enabled, the data are passed in the opposite direction. The chip therefore has two control signal inputs.

In the circuit in Figure 8-4, the memory chip data outputs are applied to a 7448 decoder, which in turn drives a seven-segment display. Alternatively, we could connect an LED to each data output pin and view their status that way.

MULTIPLEXER

We mentioned earlier in this chapter that we could use a PISO register to convert parallel data to serial data for transmission along a single line. This has to be done for transmission via a telephone line, for instance. Another way of doing this is by using a multiplexer.

Multiplexers are used in communications for simultaneous transmission of two or more signals over a common transmission medium by time division or frequency division. However, the type of multiplexer we are going to investigate here is used to select which of several data input lines is gated to a single output line, and therefore is also called a *data selector*.

The principle of this type of multiplexer is shown in Figure 8-7. This is a four-line to one-line multiplexer, in which lines A and B select which of the lines D_0, D_1, D_2, and D_3 will be connected to output line Y.

An AND gate output goes high when all inputs are high; therefore, the signal applied to the A and B inputs determines which of the AND gates will be activated if its data input is high.

Figure 8-8 shows the logic diagram of a 74151 multiplexer IC. This is an eight-line to two-line device, with three data select inputs. There is also a strobe input for pulses to enable the AND gates selected by the data select inputs.

Question: What is the purpose of the strobe line in this multiplexer? [To clock data into the system. It is therefore the same as a gating pulse.]

In practice, this IC would be used with a counter driving the data select inputs. In this way the data inputs are sequentially scanned. The same clock provides the strobe pulses.

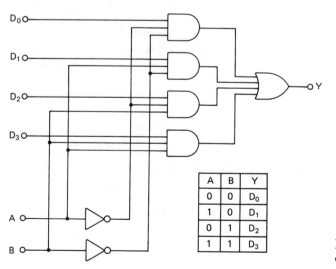

A	B	Y
0	0	D_0
1	0	D_1
0	1	D_2
1	1	D_3

Figure 8-7 Principle of a multiplexer or data selector.

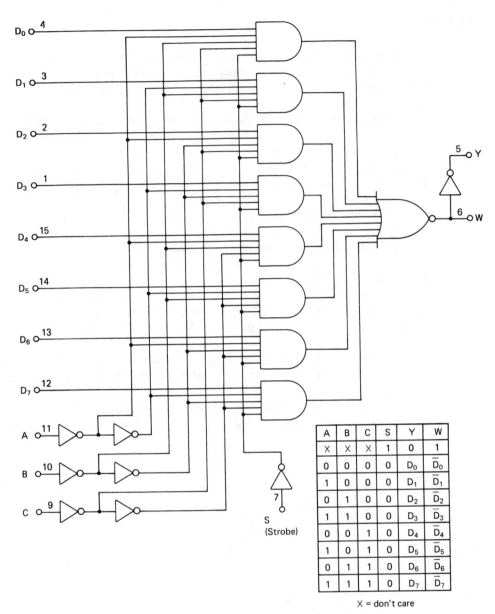

A	B	C	S	Y	W
X	X	X	1	0	1
0	0	0	0	D_0	$\overline{D_0}$
1	0	0	0	D_1	$\overline{D_1}$
0	1	0	0	D_2	$\overline{D_2}$
1	1	0	0	D_3	$\overline{D_3}$
0	0	1	0	D_4	$\overline{D_4}$
1	0	1	0	D_5	$\overline{D_5}$
0	1	1	0	D_6	$\overline{D_6}$
1	1	1	0	D_7	$\overline{D_7}$

X = don't care

Figure 8-8 Logic diagram of 74151 multiplexer.

DEMULTIPLEXER

At the other end of the single transmission line, a device is required to separate the serial signals to individual lines. Figure 8-9 shows the logic diagram of a 74138 demultiplexer, also known as a *decoder*. In this IC there are eight data outputs (pins 8 to 15) and three select inputs. There are also three enable inputs.

The select inputs select the output data line. The enable inputs are used for the data input. If $G2A$ and $G2B$ are low (they could be grounded), and if $G1$ is high, all three signals are low ($G1$ is inverted) before they are applied to the inverter inputs of the AND gate, which therefore has a high output. This is applied to all the NANDs, but only the one selected by the select inputs responds with a low output.

We can make this output high instead of low by making the $G1$ input permanently high, keeping the $G2B$ input low, and applying the data to $G2A$. Now the selected NAND is activated when $G2A$ is low, resulting in a low output. But when $G2A$ is high the NAND is deactivated, so its output is high.

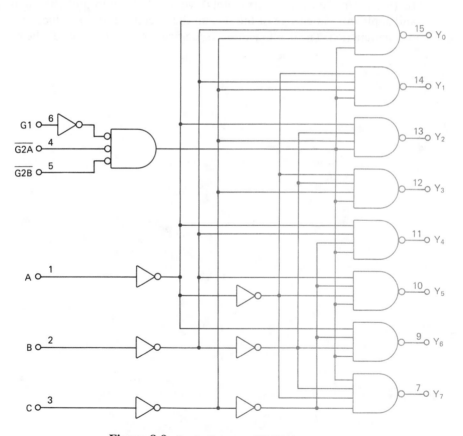

Figure 8-9 Logic diagram of 74138 demultiplexer.

The demultiplexer can also be controlled by a counter, in the same way as the multiplexer.

TROUBLESHOOTING REGISTERS

The logic probe is the most useful tool for troubleshooting logic circuits, as explained in Chapter 6. We can make a breadboard reconstruction of the circuit with an IC we know is good, and then compare the status of the pins of the good one with the suspected one. Any discrepancy will indicate the location where something is wrong.

However, if we do not have another IC of the same kind, we may be able to remove the suspected one without damaging it if it is mounted in an IC socket. First, we would list the status of all its pins when in the circuit with a problem, and then do the same when it is mounted in the breadboard circuit.

In the event that the IC is soldered in place and we do not have another "good" IC (too often the case, unfortunately), we would have to probe the pins of the IC and apply our knowledge of the IC and the circuit it is in to deduce the reason for the malfunction. The manufacturer's specifications for the IC would be very helpful.

9

Analog Circuits

Analog circuits are those in which the voltage or current output corresponds point by point to some input. The electrical signal produced by a microphone is analogous to the sound wave impinging on it. The output of a linear amplifier is an analog of its input.

Linear amplifiers are basic to most of the circuits we shall consider in this chapter. However, we shall not discuss single-transistor circuits (with which you are no doubt familiar), but will begin with the differential amplifier and go on to operational amplifiers.

DIFFERENTIAL AMPLIFIER

Amplifiers with a single transistor can achieve either high dc gain or good bias stability, but not both at the same time. This is because a large emitter resistor must be used for high bias stability, and it restricts the dc gain.

Question: What about *ac* gain? [We can bypass the emitter resistor for ac with a suitably large capacitor so that the ac signal is not degenerated.]

The problem caused by a large emitter resistor is overcome in the differential amplifier, which, as shown in Figure 9-1, employs two transistors sharing a common emitter resistor. The input is connected between their two bases so that as one is driven positive the other is driven negative. Assuming the transistors have similar operating characteristics, the emitter-current increase of one transistor is exactly

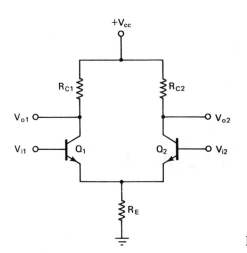

Figure 9-1 Differential amplifier.

equal to the emitter-current decrease of the other. Therefore, the total current through R_E and the voltage drop across it remain constant. Thus, R_E causes no degeneration and so requires no bypassing. An amplified output signal appears between the collectors of the two transistors.

However, if we apply the same signal or two identical signals between each base and ground simultaneously, the currents through both transistors vary simultaneously, and the resulting voltage drop across R_E opposes the input voltage, causing degeneration, as in an ordinary amplifier with an unbypassed emitter resistor.

A signal that is applied to both inputs is called a *common-mode* signal. Each output will equal the other, but since the two transistors are operating synchronously, the output between their collectors will be zero. This eliminates many undesirable signals, such as hum, when they are received by both input terminals. The ratio of differential voltage gain to common-mode voltage gain is called the common-mode rejection ratio, or CMRR.

The differential amplifier has many applications. The most widely used is in the operational amplifier.

OPERATIONAL AMPLIFIER

Figure 9-2 is a simplified drawing of the circuit of a bipolar operational amplifier. As this is an integrated circuit, there is no need to show all the details. It is divided into three basic stages. Q_1 and Q_2 are the differential amplifier, with Q_3 and Q_4 as a constant-current source. Q_5 and Q_6 are an amplifying stage. Q_7, Q_8, and Q_9 are the output stage.

An ideal op-amp should have infinite input impedance and zero output impedance. It should also have infinite *open-loop* voltage gain. Of course, such values can never be obtained in real op-amps. In the 741 op-amp, for instance, the

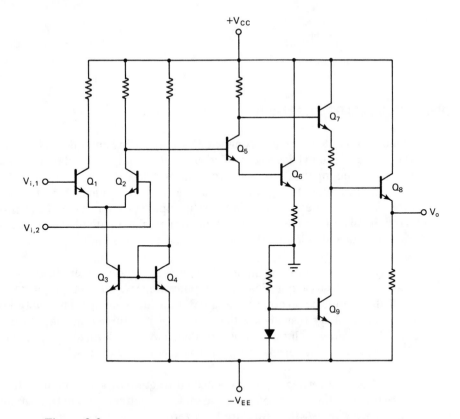

Figure 9-2 Simplified schematic diagram of a bipolar operational amplifier (Q_3 and Q_4 replace R_E in Figure 9-1).

typical values are an input resistance of 1 megohm, an output resistance of 75 ohms, and an open-loop voltage gain of 200,000. These are sufficiently close to ideal for practical purposes.

Question: Does this mean that with an input of 1 V_{dc} we can get an output of 200,000 V_{dc}? [No, because the output voltage can never exceed the supply voltage V_{cc}. See later, under *Comparator*, for some real values.]

The basic symbol for an op-amp is given in Figure 9-3. The two input terminals are designated + and −. The + terminal is called the noninverting input. When an input signal is applied to this terminal, the output signal is in phase with it. The − terminal is called the inverting input. When an input signal is applied to it, the output signal is 180 degrees out of phase with it.

Op-amps are generally connected in various feedback circuits. The gain is then called the *closed-loop* gain and is controlled by the circuit components.

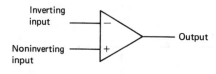

Figure 9-3 Basic symbol for op-amp (only input and output connections are shown).

INVERTING VOLTAGE AMPLIFIER

Figure 9-4 shows an op-amp connected as an inverting amplifier. This is an op-amp with feedback that yields a controlled gain and an output signal that is inverted with respect to the input. The *closed-loop gain* is designated by A_{vc}.

In calculating A_{vc}, we assume that the op-amp has ideal characteristics. Therefore, the input impedance is infinite, so the input current is zero. The open-loop gain is infinite also; so, whatever the output voltage, the input voltage, which equals the output voltage divided by the gain, must be zero as well.

> *Question:* But if the input voltage is zero, how can there be *any* output voltage? [Good question! If the input voltage were actually equal to zero, there could be no output voltage. But the input voltage is in reality a very small voltage (usually less than a millivolt), so there *is* an output voltage. However, very little error is introduced by considering the input voltage to be zero for calculating input and feedback currents.]

The noninverting input is connected to ground, so its input voltage is zero. Since the inverting input voltage is the same, the latter is said to be a *virtual ground*.

Since I_i and I_F therefore are both zero,

$$I_1 = I_F \tag{9-1}$$

which is the same as saying

$$\frac{V_i}{R_i} = \frac{-V_o}{R_F} \tag{9-2}$$

Closed-loop voltage gain is given by

$$A_{vc} = \frac{-v_o}{v_i} = \frac{-R_F}{R_1} \tag{9-3}$$

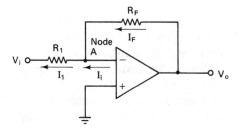

Figure 9-4 Simple inverting op-amp: $A_{vc} = -R_F/R_1$.

Since a real op-amp is a fairly good approximation of an ideal op-amp, we can use this equation to calculate the closed-loop gain of the 741 op-amp shown in Figure 9-4, where $R_F = 10$ kΩ and $R_1 = 100$ Ω:

$$A_{vc} = \frac{-10 \times 10^3}{100} = -100$$

From this we can see that it is simple to design an op-amp circuit for any voltage gain by our choice of R_1 and R_F. R_1 may be the actual internal resistance of the source, or it may be external, as in this case.

Question: What is the input impedance R_i of the op-amp in this circuit? [Because node A is at ground potential, $R_i = R_1 = 100$ Ω.]

However, it can happen that there *is* a voltage difference between the two inputs of the op-amp. This results in a dc offset voltage between them. This dc voltage is then amplified by the op-amp and appears as a larger dc voltage in the output. To eliminate this, we connect a resistor between the noninverting input and ground, with a value equal to that of the parallel combination of R_1 and R_F, assuming the two bias currents are equal. If they are not, the value of this resistor should be adjusted to compensate for their difference.

NONINVERTING VOLTAGE AMPLIFIER

If the input signal is connected to the noninverting input terminal, as shown in Figure 9-5, the output voltage will be in phase with the input. The closed-loop voltage gain is given by

$$A_{vc} = 1 + \frac{R_F}{R_1} \qquad (9\text{-}4)$$

In this example, both R_F and $R_1 = 10$ kΩ, so $A_{vc} = 2$.

As in the inverting amplifier, the voltage on the + and − inputs of a noninverting op-amp should be essentially equal, so that the voltage between them is close

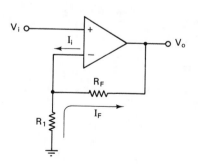

Figure 9-5   Simple noninverting op-amp: $A_{vc} = 1 + R_F/R_1$.

to zero. In this example, with $V_i = 1$ V_{p-p}, $V_o = 2$ V_{p-p}. The current flowing through R_F and R_1 is such that the voltage on the inverting input is also 1 V_{p-p}.

> *Question:* What will happen if the values of R_1 and R_F are such that the voltage on the inverting input is not approximately equal to that on the noninverting input? [The output signal will be distorted or vanish altogether, because of the unequal bias on the bases of the input transistors. In this case, a resistor of the proper value must be connected between the source and the noninverting input, as explained previously.]

BUFFER

A buffer is used to couple a low-impedance load to a source having a high internal impedance. Figure 9-6 shows how an op-amp can be used for this purpose. Since R_F and R_1 are both zero, a virtual ground exists on the inverting terminal, so $V_o = V_i$. A_{vc} is therefore 1. This circuit is also called a *voltage follower*.

VOLTAGE COMPARATOR

The voltage comparator circuit, shown in Figure 9-7, has no feedback loop, so its voltage gain is theoretically infinite. In the case of the 741, it is typically 200,000. A reference voltage is applied to the inverting input. When another voltage that is higher than the reference voltage (even if only by a few millivolts) is applied to the noninverting input, the voltage difference between the two inputs will be multiplied

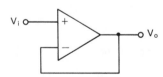

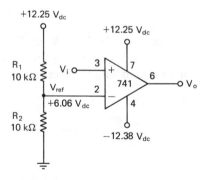

Figure 9-6 Buffer, or voltage follower, provides isolation: $A_{vc} = 1$.

Figure 9-7 Voltage comparator experiment. Measured values of supply and reference voltages shown. When $V_i = +6.4$ V_{dc}, $V_o = +11.62$ V_{dc}. When $V_i = +6.2$ V_{dc}, $V_o = -10.90$ V_{dc}. Critical $V_i = +6.25$ V_{dc}.

by some 200,000 for a very high output voltage, except that it cannot be higher than the supply voltage.

In the experiment illustrated by Figure 9-7, the reference voltage measured 6.06 V_{dc}. The positive supply voltage measured +12.25 V_{dc} and the negative −12.38 V_{dc}. The critical input voltage at the noninverting input was 6.25 V_{dc}. Increasing this voltage even slightly resulted in an output voltage of +11.62 V_{dc}; decreasing it slightly gave an output of −10.9 V_{dc}.

OFFSET COMPENSATION

Ideally, with no input voltage applied to an op-amp, the output voltage is zero. In a practical amplifier connected in the inverting or noninverting mode, however, the output is in the range of a few microvolts to millivolts for zero input. The spurious output voltage is caused by internal component differences and is called the *input offset voltage* (V_{io}). Because of the input offset voltage, we say the output contains an *error voltage*.

In theory, the dc bias currents at each of the input terminals should be equal, but in practice they are not, and their difference is the *input offset current* (I_{io}). I_{io} is typically a nanoampere current.

We can compensate for V_{io} in the 741 by installing a 10-kΩ potentiometer between pins 1 and 5. Its sliding terminal is connected to the negative supply at pin 4. The potentiometer is then adjusted for a zero output at pin 6 when there is no input signal (see Figure 9-8). Other op-amps have similar arrangements, which are explained in their manufacturers' specifications.

FREQUENCY COMPENSATION

The open-loop gain of a typical op-amp is in the region of 100 decibels (dB) at very low frequencies, but falls off to unity gain with increasing frequency. The frequency at which this occurs is called the *crossover frequency* f_c. When the op-amp is used in the closed-loop mode (which is the usual configuration), the closed-loop gain is much lower, and its response is flat to some upper frequency. This frequency may be extended, in those op-amps that have terminals for frequency compensation,

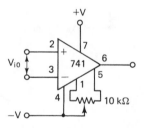

Figure 9-8 Example of offset compensation.

by connecting an *RC* circuit or a capacitor between them, as specified by the manufacturer.

DIFFERENCE AMPLIFIER

The circuit in Figure 9-9 is called a difference amplifier, or *subtractor*, because it amplifies the difference between the two inputs. The true output voltage V_o is given by

$$V_o = \frac{V_2 - V_1}{R_F/R_1} \qquad (9\text{-}5)$$

If V_1 is 200 mV and V_2 is 800 mV, V_o is given by

$$V_o = \frac{800 - 200}{100 \text{ k}\Omega/100 \text{ k}\Omega} = 600 \text{ mV}$$

SUMMING AMPLIFIER

While we usually do this and the following arithmetical functions with a pocket calculator or computer, these circuits illustrate some other ways in which the versatile op-amp can perform.

Just as it is possible to use an op-amp for amplifying the difference between two voltages, we can also use it to amplify the sum of two or more voltages. Figure 9-10 shows that in the summing circuit the noninverting input is grounded. As in the inverting amplifier, the inverting input is a virtual ground. When the voltages to be added are applied to it, their total is amplified and appears at the output.

If we make the input resistances R_1, R_2, and so on, equal to each other and call this value R, the output voltage V_o is given by

$$V_o = -(V_1 + V_2 + \cdots + V_N)\frac{R_F}{R} \qquad (9\text{-}6)$$

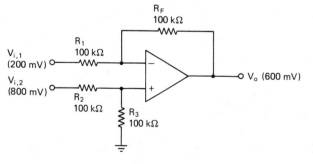

Figure 9-9 Difference amplifier.

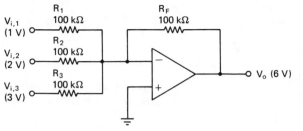

Figure 9-10 Summing amplifier.

Question: How do we sum voltages that add to a total value higher than the maximum V_o the op-amp can achieve? [We have to change the ratio R_F/R to reduce the total input to a level that will give a V_o below the maximum value and then afterward multiply V_o by this ratio to get the proper total.

If V_1, V_2, and V_3 had been 100, 200, and 300 V_{dc}, respectively, we would have had to make R_1, R_2, and R_3 1 MΩ each and R_F 10 kΩ. This would give us $R_F/R = 0.01$. V_o would still have been -6 V$_{dc}$, but dividing it by 0.01 would get back its true value of -600 V$_{dc}$.]

LOGARITHMIC AMPLIFIER

If we want to use an op-amp for multiplication, division, exponentiation, or extraction of roots, we must first obtain logarithms of the numbers involved. Then we can add or subtract the latter and get the antilogarithm of the result. A diode or a transistor in the feedback loop provides the logarithmic characteristic. In Figure 9-11, a transistor is used. Since the output is temperature dependent, a thermistor and resistor R_1 have been added to compensate. (The thermistor should have a temperature coefficient of 0.35%/°C, and the value of R_1 should be 15 times the cold resistance of the thermistor.)

The output voltage V_o is given by

$$V_o = -26 \text{ mV} \frac{\ln V_i}{R - \ln I_{ES}} \tag{9-7}$$

where I_{ES} is the base–emitter reverse saturation current. However, if you prefer to work in common logarithms, the -26-mV term should be changed to -60 mV.

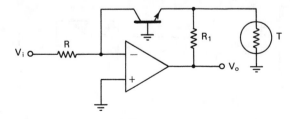

Figure 9-11 Logarithmic amplifier. R_1 and T are for temperature compensation.

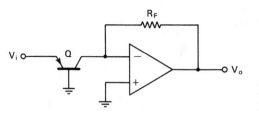

Figure 9-12   Antilog amplifier (temperature compensation not shown).

ANTILOG AMPLIFIER

The circuit in Figure 9-11 becomes an antilog amplifier by moving the feedback transistor and temperature-compensation circuit to the input, as shown in Figure 9-12. For an input $V_i = \ln z$,

$$V_o = -I_{ES} R_F \, \epsilon^{39 \, \epsilon \ln z} \tag{9-8}$$

VOLTAGE MULTIPLIER AND DIVIDER

Figure 9-13(a) shows how the four circuits just described can be combined to provide a multiplier–divider, and Figure 9-13(b) is a practical realization of such a concept. Note that the thermistors have been replaced by diodes and that the source voltages have not been shown.

The circuit can multiply two numbers, divide one number by another, or multiply two numbers and divide their product by a third. The output voltage is given by

$$V_o = \frac{V_{i1} V_{i3}}{10 V_{i2}} \tag{9-9}$$

Question: Are op-amps used in pocket calculators? [No. Pocket calculators use binary logic circuits for addition and subtraction and repeated additions or subtractions for multiplication or division.]

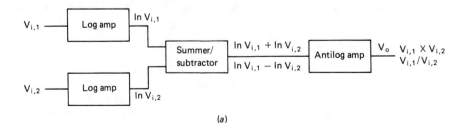

(a)

Figure 9-13(a)   Multiplier–divider block diagram.

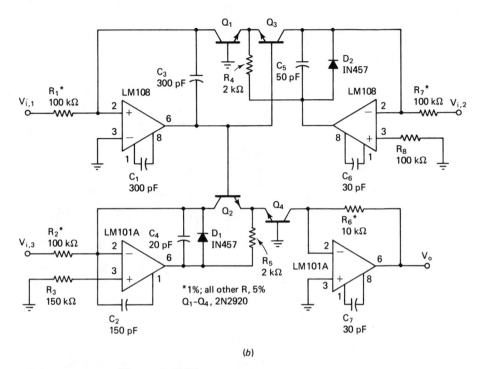

Figure 9-13(b) Practical multiplier–divider circuit.

EXPONENTIATION AND EXTRACTION OF ROOTS

Raising numbers to a power and finding the root of a number are simple operations using logarithms. To raise a number to a power, we multiply the logarithm of the number by the value of the exponent and take the antilogarithm of the result. To find the root of a number, we divide the logarithm of the number by the value of the root and take the antilogarithm of the result. The circuit of Figure 9-13(b) could be modified to do this.

WAVE GENERATION AND SHAPING

Op-amps are widely used in the generation of square and sinusoidal waves, and also of pulses and other nonsinusoidal functions.

SQUARE-WAVE GENERATOR

Figure 9-14 shows a free-running multivibrator. C_1 and R_4 determine the period of oscillation T:

$$T = 1.388(R_4 C_1) \tag{9-10}$$

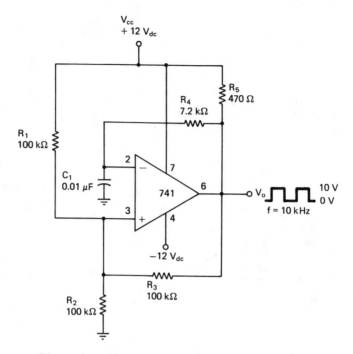

Figure 9-14   Square-wave generator with a 741 op-amp.

Substituting values from Figure 9-14,

$$T = 1.388(7.2 \times 10^3 \times 0.01 \times 10^{-6})$$

$$= 100 \ \mu s$$

Since $f = 1/T$, the frequency must be 10 kHz.

SINUSOIDAL OSCILLATOR

Figure 9-15 shows a Wien-bridge oscillator. The gain of the amplifier is controlled by R_3 and R_4. The frequency of oscillation is given by

$$f = \frac{0.159}{RC} \tag{9-11}$$

where $R = R_1 = R_2$ and $C = C_1 = C_2$.

PULSE GENERATOR

Figure 9-16 shows a pulse generator. It is essentially the same as the square-wave generator of Figure 9-14. However, the feedback circuit now consists of two diodes

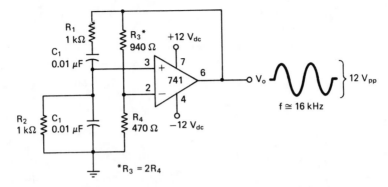

Figure 9-15 Wien–bridge oscillator with a 741 op-amp.

and two variable resistors. R_4 is adjusted to give the *on* time t_1, and R_5 is adjusted for the *off* time t_2.

$$t_1 = 0.694(R_4 C_1) \tag{9-12}$$

$$t_2 = 0.694(R_5 C_1) \tag{9-13}$$

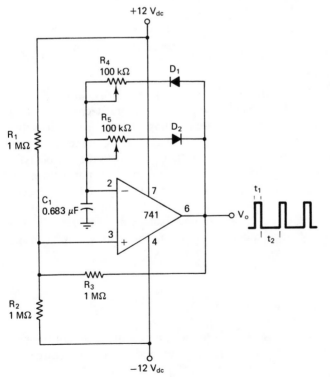

Figure 9-16 Pulse generator.

FUNCTION GENERATORS

The preceding circuits can be combined to give function generators (see Chapter 3), but as function-generator ICs are available with the circuits provided internally, it is not necessary.

ACTIVE FILTERS

Filter circuits using op-amps are very popular since no inductors are required. They can be used for the frequency range from dc to 500 kHz. *LC* filters, which cover the range from 100 Hz to 300 MHz, require large and expensive inductors at low frequencies and do not have the flexibility of active filters, which, if required, can be provided with various values of input and output impedance, as well as voltage gain.

If you decide to build any of the following filter circuits, your results may not be exactly the same. This is to be expected due to variations in real component values, as opposed to nominal values.

Question: What other advantages does an active filter have compared to a passive filter? [It has gain, which can be obtained in any desired amount by the choice of the circuit components.]

LOW-PASS FILTER

Figure 9-17 shows a simple low-pass filter. The feedback resistor is paralleled by a capacitor. As the frequency increases, this capacitor's reactance decreases so that more feedback opposes the input signal. The output voltage decreases until its value is 3 dB down from the input voltage. This is called the cutoff frequency f_c.

In a simple filter such as this, the rolloff is rather gradual (6 dB per octave). Figure 9-18 shows a better low-pass filter that has a rolloff of 12 dB per octave. An additional capacitor in the input allows a portion of the input signal to be grounded. This portion increases with frequency so that less of the signal is applied to the input of the op-amp. Together with the capacitor in the feedback circuit, this results in a steeper dropoff of the output voltage gain as the frequency increases, as we can see from the response curve.

HIGH-PASS FILTER

Figure 9-19 shows a simple high-pass filter. It is the same as the low-pass filter in Figure 9-17, but with capacitors where the resistors were, and vice versa. Figure 9-20 shows a better high-pass filter with improved rolloff.

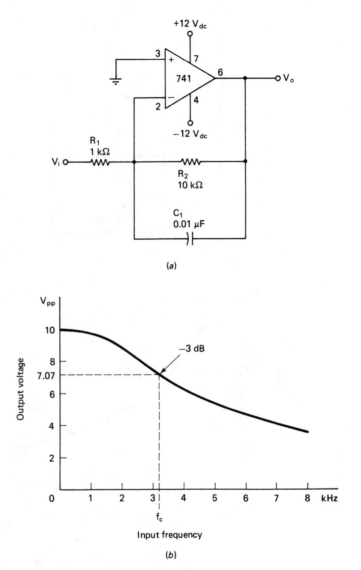

Figure 9-17 Simple low-pass active filter: (a) circuit; (b) response curve.

BANDPASS FILTER

Bandpass filters are either wide band or narrow band. A bandpass filter is **narrow** band if the ratio of the upper 3-dB frequency to the lower 3-dB frequency is 1.5, or less.

A wide-band bandpass filter is achieved by cascading a high-pass filter **and a**

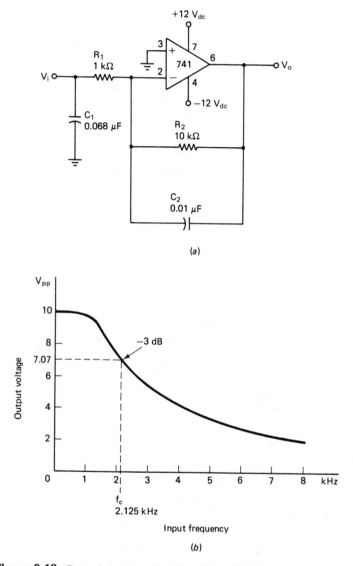

Figure 9-18 Better low-pass active filter: (a) circuit; (b) response curve.

low-pass filter. The high-pass filter is designed to attenuate frequencies lower than the passband, and the low-pass filter attenuates those higher than the passband.

A narrow bandpass filter uses the circuit shown in Figure 9-21. This circuit is resonant at the center frequency, and it is a good idea to make R_2 adjustable for fine tuning.

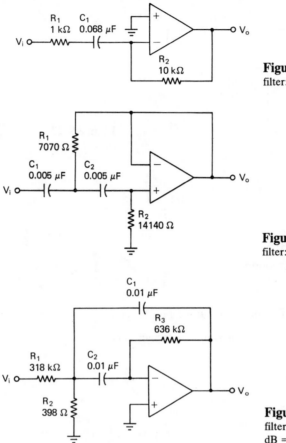

Figure 9-19 Simple high-pass active filter: $f_c = 2.7$ kHz.

Figure 9-20 Better high-pass active filter: $f_c = 3$ kHz.

Figure 9-21 Narrow bandpass active filter: $f_c = 1000$ Hz; bandwidth at -3 dB $= 50$ Hz.

BAND-REJECT OR NOTCH FILTER

Wide-band band-reject active filters have a ratio of upper 3-dB frequency to lower 3-dB frequency of 1.5 or more, and are designed by paralleling a low-pass and a high-pass filter, as shown in Figure 9-22. The outputs of the two filters are then combined by using a third op-amp.

A narrow-band notch active filter uses a twin-T network, as shown in Figure 9-23. This gives a deep null at the center frequency. The component values are calculated from

$$Q = \frac{F}{BW} \tag{9-14}$$

$$R_1 = \frac{1}{2\pi FC} \tag{9-15}$$

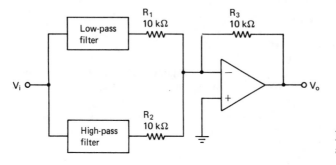

Figure 9-22 Wide-band band-reject active filter.

$$K = \frac{4Q - 1}{4Q} \qquad (9\text{-}16)$$

where Q = circuit quality factor

$\qquad F$ = center frequency

$\qquad BW$ = 3-dB bandwidth

$\qquad R$ = arbitrarily chosen value of resistance

$\qquad C$ = arbitrarily chosen value of capacitance

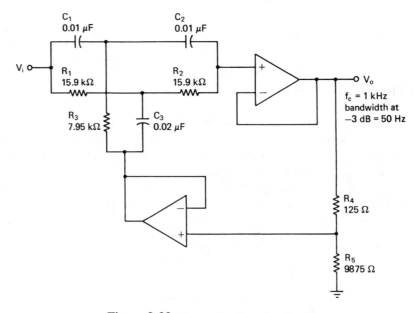

Figure 9-23 Narrow-band notch active filter.

In this example, with $R = 10$ kHz and $C = 0.01$ μF,

$$Q = \frac{1000}{50} = 20$$

$$R_1 = \frac{1}{2 \times \pi \times 1000 \times 0.01 \times 10^{-6}} = 15.9 \text{ k}\Omega$$

$$R_2 = R_1$$

$$R_3 = \frac{R_1}{2} = \frac{15.9 \times 10^3}{2} = 7.95 \text{ k}\Omega$$

$$K = \frac{4 \times 20 - 1}{4 \times 20} = 0.9875$$

$$R_4 = (1 - K)R = (1 - 0.9875) \times 10 \times 10^3 = 125 \ \Omega$$

$$R_5 = KR = 0.9875 \times 10 \times 10^3 = 9.875 \text{ k}\Omega$$

$$C_1 = C_2 = C = 0.01 \ \mu\text{F}$$

$$C_3 = 2C = 0.02 \ \mu\text{F}$$

TROUBLESHOOTING OP-AMPS

Troubleshooting an op-amp circuit is usually simple. There is very little to check: voltages, dc and ac, at the input; power supply; output pins; and perhaps the offset compensation pins, if used.

For example, consider a situation where there is no output from a 741 op-amp connected as an inverting amplifier, as in Figure 9-4. The two power supplies are correct, and a sine-wave signal appears at input pin 2. So far, so good. But when we check the dc voltage at this pin, we find $5.0 \ V_{dc}$ instead of the virtual ground that should be there. This voltage must be biasing the input differential amplifier to cutoff. We observe that there is a coupling capacitor between the input resistor and the signal source and assume it must be shorted. When we replace this capacitor, normal operation is restored.

Question: What would happen if this capacitor were open instead of shorted? [The dc voltage at pin 2 would be correct, but there would be no ac signal present and, consequently, no output signal.]

10

Analog–Digital and Digital–Analog Conversion

We can easily process, store, transmit, and display information in digital form, without error or loss. We have available many low-cost devices that enable us to better handle analog variables, such as voltage, temperature, sound, and so on, by first converting them to digital form.

In some cases, we don't need to go any further. A digital voltmeter, for instance, after processing the input analog voltage, displays it as a number, which is all we want it to do. However, we can also change the digital data back into analog form, if we so desire, for other types of display or control of real-world variables. The means for doing this are the analog-to-digital (A/D) converter and the digital-to-analog (D/A) converter.

A/D CONVERSION

A/D conversion has two stages. We call the first stage *quantizing*, the second *encoding*. These are illustrated by the circuit in Figure 10-1. In this circuit there are three voltage comparators. In Chapter 9 we saw how a voltage comparator has a fixed reference voltage at one input and the voltage to be compared on the other. Now we have three voltage comparators, each with a different reference voltage applied to its inverting input terminal. Since the maximum reference voltage is $+12 \text{ V}_{dc}$ and each resistor has the same value, the voltage divider applies $+9 \text{ V}_{dc}$ to comparator 1, $+6 \text{ V}_{dc}$ to comparator 2, and $+3 \text{ V}_{dc}$ to comparator 3.

An analog voltage that has a range from 0 to $+12 \text{ V}_{dc}$ is applied simultaneously to the noninverting inputs of all three comparators. If its value is less than $+3 \text{ V}_{dc}$,

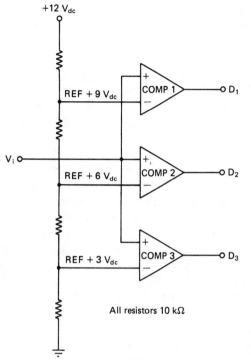

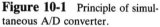

Figure 10-1 Principle of simultaneous A/D converter.

there will be no change in the output of any comparator. They will all be low. But if its value is between +3 and +6 V_{dc}, the output of comparator 3 will go high.

Similarly, if the analog voltage is between +6 and +9 V_{dc}, the outputs of both comparator 3 and comparator 2 will go high. And if the analog voltage is over +9 V_{dc}, all three comparators will go high.

In this way, we have partitioned the analog voltage range into four quanta and have encoded its actual value into a 3-bit word, as in Table 10-1. This is a simple example of a *simultaneous* A/D converter. It may also be called a flash or parallel converter. It is obvious that using only three comparators results in very coarse resolution; so, although it has the advantage of being very fast, good resolution would require a much larger number of comparators.

TABLE 10-1 QUANTIZATION AND ENCODING EXPLAINED

Input Voltage (V_{dc})	Encoded Output		
0 to +3	0	0	0
+3 to +6	0	0	1
+6 to +9	0	1	1
+9 to +12	1	1	1

Question: How many comparators would be required for eight digital output lines? [The number is given by $2^n - 1$, where n is the number of digital output lines. In this case, the number would be $2^8 - 1$, or 255 comparators.]

We can overcome this disadvantage by using a variable reference instead of a fixed one. Then we only need one comparator. The variable reference is obtained from a counter, as shown in Figure 10-2. The binary output of the counter can be converted into a staircase ramp by means of a D/A converter. (We'll see how the D/A converter works later.)

This staircase ramp is the reference voltage and is applied to the inverting input of the comparator. As long as the reference voltage is less than the analog input voltage at the noninverting input, the comparator output will be high; so it and the positive-going clock pulses at the other input of the AND gate will continue to enable the gate, and its output of positive pulses will drive the counter.

But when the reference voltage exceeds the analog voltage by even a few millivolts, the comparator output will go low, and the AND gate will no longer respond to the clock pulses. This stops the counter, and it displays the count. The digital display actually will be in the form of a reading of whatever quantity is being measured, in this case, volts.

An improved version of this circuit employs *dual-slope conversion*, shown in Figure 10-3. In this circuit the analog input is first applied to an integrator. The analog input is integrated for a fixed number of clock pulses, as counted by the counter. At the end of this time, the integrator input is switched to a reference voltage of opposite polarity so that the integrator's output returns to zero. The counter counts the number of clock pulses during this second slope. This number of pulses is variable, depending on how high the first ramp got, and is therefore proportional to the input voltage.

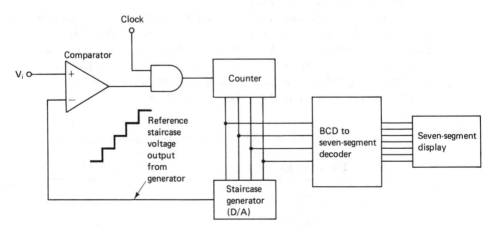

Figure 10-2   A/D converter using a counter to generate a staircase reference voltage.

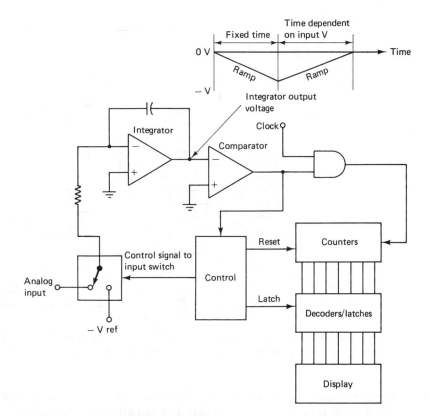

Figure 10-3 Dual-slope A/D converter used in DMMs.

This type of A/D converter is suitable for digital voltmeters and the like, but it is still too slow for computer interface systems. For these, a *successive-approximations* converter is used. A block diagram of this converter is shown in Figure 10-4.

For simplicity, let us assume that the output word of the converter has 4 bits. Then the reference voltage V is quantized into four: $V/2$, $V/4$, $V/8$, and $V/16$, corresponding to the binary words 1 0 0 0, 0 1 0 0, 0 0 1 0, and 0 0 0 1. When the analog voltage to be digitized is first applied to the comparator, the $V/2$ value is used for the reference. If the analog voltage is higher than this, the comparator output is high and a 1 is stored in the shift register. The $V/2$ reference voltage remains on the inverting input. If the analog input had been less than $V/2$, the comparator output would have gone low and a 0 would have been stored in the shift register. The $V/2$ reference voltage would have been removed from the inverting input.

Next, the $V/4$ reference voltage is applied to the inverting input. If the $V/2$ reference is also there, the new reference voltage will be $3V/4$, but if it is not, the new reference will be $V/4$. If the analog voltage is higher than the $3V/4$ reference voltage, then the comparator output again will be high, and the next bit stored in

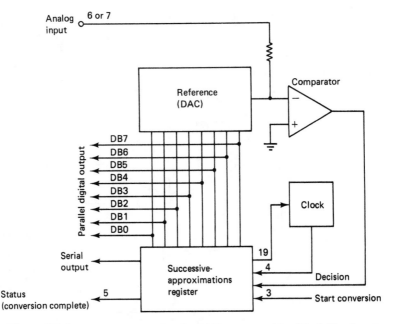

Figure 10-4 Successive-approximations A/D converter (simplified). Numbered inputs and outputs are the corresponding pin numbers of the ADC0804 given in Table 10-2.

the shift register will be a 1 and the $3V/4$ reference voltage will remain. But if it is not, then the comparator output will be low and a 0 will be stored. The $V/4$ reference voltage will be removed.

Question: This seems rather complicated. Isn't there an easier way of explaining it? [This could be compared to weighing something in a balance, where you have a set of weights with values of $\frac{1}{2}$, $\frac{1}{4}$, $\frac{1}{8}$, and $\frac{1}{16}$ pound. As shown in Figure 10-5, when weighing the mass in the left pan, we first put the $\frac{1}{2}$-lb weight in the right pan. If it does not cause the pan to go down, we add the $\frac{1}{4}$-lb weight; but if the $\frac{1}{2}$-lb weight does make the pan go down, we remove it and substitute the $\frac{1}{4}$-lb weight. If adding the $\frac{1}{4}$-lb weight to the $\frac{1}{2}$-lb weight makes the right pan go down, then we remove the $\frac{1}{4}$-lb weight and substitute the $\frac{1}{8}$-lb weight. If this does not make the pan go down, we add the $\frac{1}{16}$-lb weight. If the pan now balances, we have determined the weight of the mass to be $\frac{1}{2} + \frac{1}{8} + \frac{1}{16}$ lb, or 11 ounces.]

By quantizing the reference into 4 quanta we get 16 possible values from 0 0 0 0 to 1 1 1 1. Put otherwise, our output word of 4 bits gives us 2^4 discrete values. If we had an output word of 8 bits, it would give us 2^8, or 256 discrete values, and so on, for larger words.

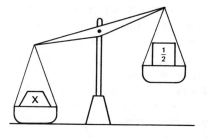

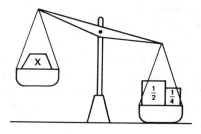

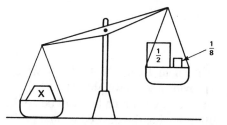

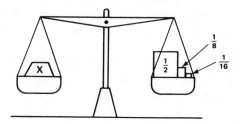

Figure 10-5 Weighing X by successive approximations.

A successive-approximations converter can combine useful resolution, up to and beyond 12 bits, with a fairly short conversion time (less than 12 μs for 12-bit conversion). A further advantage is that conversion time is fixed and independent of the magnitude of the analog input, permitting efficient interfacing with computers.

Another area of use for A/D converters, where speed is essential, is in digital audio systems. In these, the analog audio must be changed to digital and then back to audio before being applied to the audio output amplifier stage. The audio signal is sampled, and the samples are digitized. Later the digitized samples are changed back into analog. If the sampling is slow, it will not be possible for the reconstituted analog signal to have the smoothness of the original. It will have coarse serrations that will affect the quality of the sound. Fast conversion will greatly increase the sampling rate, and the much finer serrations will be imperceptible and can probably be removed altogether with a suitable filter capacitor.

The successive-approximations converter is exemplified in the ADC0804 integrated circuit by National Semiconductor. This MOS device is designed to interface with the 8080 microprocessor (although it has other users also). The functions of its 20 pins are described in Table 10-2.

The data output pins of the ADC0804 are designed to sink 16 mA and source 6 mA each. When interfacing with a microprocessor, this has no significance; but if we want to interface with a device that draws current, we must consider whether the required current exceeds 6 mA for a high output. An example would be an LED connected between the data output and ground. If the current it draws (with a series resistor, of course) is less than 6 mA, it will light on a high. But it cannot light if it requires more current.

D/A CONVERSION

A simple 6-bit D/A circuit is shown in Figure 10-6. Its basic elements are the reference; a resistive network to provide a set of weighted (32, 16, 8, 4, 2, 1) voltages, currents, or gains; a set of switches to select which bits will be applied to the input; and an op-amp to provide an output having the desired format (voltage or current), level, and impedance. Notice its similarity to the summing amplifier in Chapter 9.

When a switch is closed, $-V_{REF}$ appears across the resistor connected to it, because the summing point is at virtual ground. This causes a current V_{REF}/R_i to flow from the summing point through the resistor, switch, and reference back to ground. The same happens in each switch circuit when its switch is closed. The only way this current, or currents if more than one switch is closed, can be provided is from the output of the op-amp through feedback resistor R_F. Therefore, the output voltage V_o has to be of the proper value to give

$$\frac{V_o}{R_F} = \frac{V_{REF}}{R_1} + \frac{V_{REF}}{R_2} + \cdots + \frac{V_{REF}}{R_n} \qquad (10\text{-}1)$$

Question: Is this circuit used in practice? [No. The wide range of resistances required is difficult to provide economically, and the switches give rise to transient spikes. Furthermore, the load on the summing point and the reference is too variable. Modern D/A converters avoid this by using *resistive-ladder networks*.]

TABLE 10-2 PINS AND FUNCTIONS OF THE ADC0804 CONVERTER

Pin No.	Function	Explanation of Function
1	Chip select $\overline{CS}$	Connects to the microprocessor or to ground (must be low for converter to operate).
2	Read $\overline{RD}$	Connects to the microprocessor or to ground (must be low for the digital output to be read).
3	Write $\overline{WR}$	Start conversion (must be low for conversion to start). (On some devices it is designated $\overline{SOC}$).
4	Clock in CLK IN	Input for clock signal from internal clock (pin 19) or external clock.
5	Interrupt $\overline{INTR}$	Goes low to signal end of conversion ($\overline{EOC}$ on some devices). If connected to pin 3, conversion repeats immediately.
6 7	$V_{IN(+)}$ $V_{IN(-)}$	Analog inputs. Usually the + input is used, with the − input grounded. Both may be used as in a differential amplifier to reject common-mode input.
8	Analog gnd A GND	Analog input ground.
9	$V_{ref/2}$	Reference voltage input. If left unconnected, V_{ref} is +5 V_{dc}. V_{ref} is also +5 V_{dc} if +2.5 V_{dc} is applied to pin 9. Other V_{ref} ranges may be obtained by applying other voltages. (A voltage may also have to be applied to pin 7 to absorb the offset.)
10	Data gnd D GND	Digital output ground.
11–18	Data out DB7–DB0	Buffered data outputs.
19	Clock out CLK R	Output from internal clock. Frequency determined by value of resistor (typically 10 kΩ) connected between pin 19 and pin 4, and capacitor from pin 4 to ground: $f_c = 1/1.1RC$.
20	V_{cc}	Typically +5 V_{dc}, maximum +6.5 V_{dc}.

RESISTIVE-LADDER NETWORK

A simplified version of a resistive-ladder network is shown in Figure 10-7(a). There are only four digital inputs. Let us begin by grounding all four of them. At node 1 the two 20-kΩ resistors are parallel to each other; therefore, their combined resistance is 10 kΩ. We can replace these two resistors by a single 10-kΩ resistor, as shown in Figure 10-7(b).

If we now go to node 2, we see that we have the same situation. The resistance

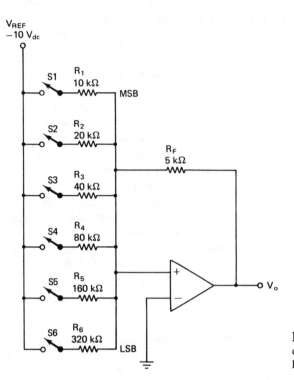

V_{REF}
−10 V_{dc}

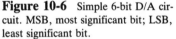

Figure 10-6 Simple 6-bit D/A circuit. MSB, most significant bit; LSB, least significant bit.

from node 2 to ground in either direction is 20 kΩ, and these resistances are in parallel; therefore, they can be replaced also by a single 10-kΩ resistor.

The same occurs at each of the remaining nodes, 3 and 4, from which we see that the resistance from any node to ground or input is 20 kΩ, and the resistance from any node through both input and ground is 10 kΩ. We could use other values of resistance, but always in the ratio of $2R$ and R.

Now suppose we apply 5 V_{dc} to the left input (node 4). This is the most significant bit, with a decimal weight of 8. The resistance at node 4, as we have just seen, is the same (20 kΩ) to both input and ground. The voltage at node 4 must therefore be $\frac{5}{2}$, or 2.5 V_{dc}. In more general terms, we call it $V/2$. This voltage appears at the analog output.

Now let's apply 5 V_{dc} to the next input to the right (node 3), which has a decimal weight of 4. The situation can be represented by the equivalent circuit of Figure 10-7(c). The portion of the binary ladder to the right of node 3 can be replaced by a single 20-kΩ resistor (since these inputs are at ground potential). Therefore, the voltage at node 3 is $V/2$. This now becomes the input to node 4. The actual resistance at node 4 to ground or input is 20 kΩ, so the voltage at node 4 (also the analog output) must be $V/4$.

Following the same line of reasoning, we see that the voltage at the next input to the right (node 2) is $V/8$, and the voltage of the least significant bit (node 1) is $V/16$. If all digital inputs were high, the analog output voltage would be

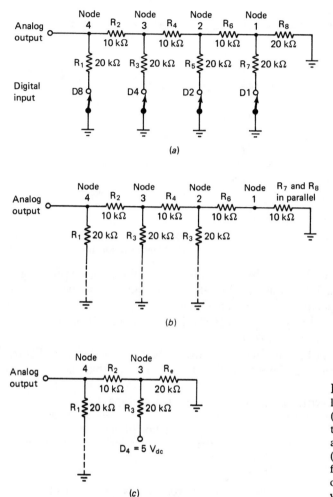

(a)

(b)

Node
4 R_2 Node
3 R_e

Analog output

10 kΩ   20 kΩ

R_1 20 kΩ R_3 20 kΩ

$D_4 = 5\ V_{dc}$

(c)

Figure 10-7 Principle of resistive-ladder network: (a) simplified version; (b) same version, with equivalent resistance of R_7 and R_8 in parallel; (c) equivalent circuit with input at D_4 (R_e = equivalent resistance to ground from node 3). Applying Thevenin's theorem, voltage at node 3 = $V/2$, and voltage at node 4 = $V/4$.

$$\frac{V}{2} + \frac{V}{4} + \frac{V}{8} + \frac{V}{16} \quad \text{or} \quad 2.5 + 1.25 + 0.625 + 0.3125$$

giving a total of 4.6875 V, which is less than the source voltage of 5 V. If we used a ladder with eight inputs, the difference would be less (0.0195 V); but however small the increment, we can never reach 5 V, only approach it.

Therefore, the resolution of this converter is equal to the smallest voltage increment that can be provided in the analog output. This, in turn, depends on the number of digital inputs and is given by $V/2^n$, where n = number of digital inputs. As we have just seen, a 4-bit input gives an increment of $V/16$, so an 8-bit input would give one of $V/256$, and an 18-bit input would give an increment of $V/262,144$.

We normally use D/A converters that are in the form of an IC, of which many different types are available. Most have an 8-bit input, although inputs up to 18 bits

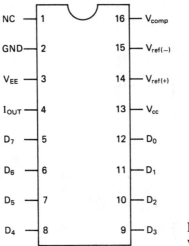

Pin	Signal	Pin	Signal
NC	1	16	V_{comp}
GND	2	15	$V_{ref(-)}$
V_{EE}	3	14	$V_{ref(+)}$
I_{OUT}	4	13	V_{cc}
D_7	5	12	D_0
D_6	6	11	D_1
D_5	7	10	D_2
D_4	8	9	D_3

Figure 10-8   MC1408-8N D/A converter pins and functions.

are found in some. The accuracy of the converter is determined by the precision of the voltages or currents and the resistor values used in the ladder. This will be stated in the manufacturer's specification as a percentage.

A typical D/A converter is the MC1408-8N shown in Figure 10-8. Its pins and their functions are explained in Table 10-3.

The MC1408-8N can be interfaced with the ADC0804, as shown in Figures 10-9(a) and 10-9(b). This is a simplified digital audio system that illustrates the

TABLE 10-3   PINS AND FUNCTIONS OF THE MC1408-8N D/A CONVERTER

Pin No.	Function	Explanation of Function
1	NC	No connection
2	GND	System ground
3	V_{EE}	Negative supply voltage (-4.5 to -12 V_{dc})
4	I_{OUT}	Output current
5–12	$D7$–$D0$	Digital inputs
13	V_{CC}	Positive supply voltage (4.5 to 12 V_{dc})
14	$V_{ref(+)}$	Positive reference voltage used internally for the ladder. If a negative reference is desired, pin 14 is grounded via a 1- to 2.5-kΩ resistance, and the negative reference voltage is applied to pin 15.
15	$V_{ref(-)}$	Negative reference voltage (usually grounded in same way as pin 14 when a positive reference is used).
16	V_{comp}	Connected via a capacitor to pin 3 for phase-shift compensation.

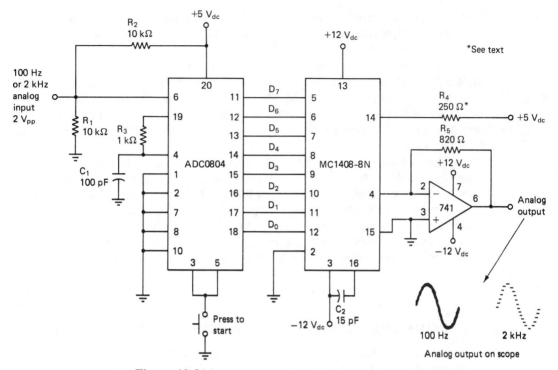

Figure 10-9(a)   Experiment to show principle of digital audio system.

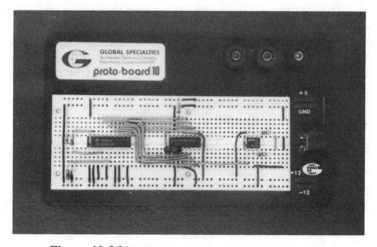

Figure 10-9(b)   Breadboard realization of Figure 10-9(a).

principle of operation of the more advanced versions used in modern audio disk and audio tape equipment. In practice, additional stages, such as amplifiers, would be between the A/D converter and the D/A converter.

When we apply a sine wave to the analog input (pin 6) of the ADC0804, it is digitized. We then apply the digital output (pins 11 to 18 of the ADC0804) to the digital input (pins 5 to 12) of the MC1408-8N. The analog output of the D/A converter (pin 4) is then applied to the inverting input of a 741 op-amp to bring it into phase with the analog input at pin 6 of the ADC0804 (the MC1408-8N has a negative output voltage).

When we apply a 100-Hz sine wave to the analog input, we get a fairly good reproduction of it at the analog output. On the oscilloscope, we can see the individual samples of the original input signal, but it doesn't sound too bad in an earphone.

However, if we apply a higher frequency, say, 2 kHz, the sampling becomes much coarser, and we can hear a definite buzz in the earphone. This is because the ratio of input frequency to sampling frequency is less. Of course, if the A/D and D/A converters also had greater resolution it would make a considerable difference as well.

TROUBLESHOOTING A/D AND D/A CONVERTERS

Like most ICs, an A/D converter either works or it doesn't. However, it may have an output with an offset. This should be investigated and a correcting voltage applied to pin 7, if necessary. Figure 10-9 shows a test setup that will allow us to verify the operation of an ADC0804 or similar IC.

Troubleshooting these circuits requires that we understand the effects of both digital and analog problems. In the circuit of Figure 10-9, we can simulate such problems by disconnecting one of the pins or changing the value of a component.

For instance, disconnecting any one of the digital input or output pins will severely distort the output sine wave. The sound will be terrible, and its appearance on the oscilloscope will be terrible also.

If we remove the resistor connecting pin 14 of the D/A to the +5 V_{dc} supply (as if it had opened up), we will get no output at all.

In other words, if we reproduce the circuit on a breadboard, we can do things to it that will lead us to where the problem is in the circuit we are troubleshooting.

> *Question:* What is the purpose of V_{ref} at pin 14? [It enables us to produce a controlled output voltage by setting the input current reference. If we were to use a potentiometer instead of a fixed resistor, we could set this exactly.]

Appendix A

Semiconductors

The most widely used semiconductor is *silicon*. Unlike metals, in which free electrons are abundant, the electrons in its outer shell are not so easily liberated. However, if they can acquire sufficient energy from heat or other external stimulus, they may break loose temporarily. The atom then has a deficiency of one negative charge or, to put it another way, a surplus of one positive charge. It is not long before this positive charge recaptures the truant electron, or another one, and things revert to normal. However, while the electron was gone, there was a "hole," or empty space in the atom.

If the positive terminal of a battery is connected to one side of a piece of silicon and its negative terminal to the other side, any electrons that can get free will move in the direction of the positive terminal. As electrons leave their atoms and move away, they leave holes behind. These holes capture following electrons and disappear. However, the following electrons leave holes behind, too; therefore, as the electrons go hopping from atom to atom toward the positive terminal, it seems as if the holes are drifting in the opposite direction.

Since these electrons are jumping from atom to atom, they are not the same as the free electrons in a metal conductor, which do not leave holes behind. Their movement is much slower. We usually consider them as hole current, rather than electron current.

However, the flow changes dramatically when very small quantities of certain other elements are added to the silicon. These additives are called *impurities*.

The electrons that can be detached in a silicon atom are the four outermost ones. They are farthest from the nucleus; therefore, the binding force that holds the atom together is weakest for them. If semiconductor atoms had as many electrons as they would like, they would have eight in this outermost *shell*. As they cannot, each attaches itself to four neighboring semiconductor atoms so it can share their four outermost electrons, and, of course, the same goes for the other atoms also. This is called *covalent bonding*, and the electrons concerned are called *valence electrons*. Covalent bonding causes the atoms to lock together in a *crystal lattice*.

If impurity atoms with five valence electrons are added, they lock into the crystal lattice also; but since each atom needs only four electrons to do this, a spare electron results. This electron is a free electron, similar to a free electron in a conductor.

Silicon with free electrons like this is called *n type*, because electrons are negative. However, it does not have an overall negative charge, because the impurity atoms from which they came now each have a positive charge.

Impurity atoms with three outermost electrons are also used to *dope* silicon to make it *p type*. As you might expect, they produce holes; but since the holes are really gaps in the crystal lattice, they are not free like the surplus electrons in *n*-type silicon. Nor do they confer an overall positive charge, because the total numbers of electrons and protons (positive charges in the nucleus) are still equal.

Appendix B

Diodes and Transistors

In a semiconductor diode there is a junction of n-type silicon and p-type silicon, as shown in Figure A-1. At first, free holes (black) from the p-type region diffuse across the junction into the n-type region, and free electrons (white) diffuse across the junction in the opposite direction. The p-type region is losing holes and gaining electrons, acquiring a negative charge. The n-type region is losing electrons and gaining holes, acquiring a positive charge. The negative charge on the p-type silicon repels electrons from the vicinity of the junction and prevents any more from crossing it; and the positive charge on the n-type silicon repels holes from the vicinity of the junction and prevents any more from crossing it. Consequently, the vicinity of the junction is called the *depletion layer*. If an external battery is connected to the diode as shown, its voltage adds to the potential difference between the two sides of the depletion layer, so that it is made even more effective in blocking current flow.

Only a very small current due to *minority carriers* (electrons and holes created by thermal agitation) flows across the junction. However, if the polarity of the applied voltage is reversed, the *majority carriers* (those created by the impurity atoms) flow across the junction and become a current flowing from the negative terminal of the battery through the n-type region, across the junction, through the p-type region, and back to the positive terminal of the battery.

In other words, the junction allows current to flow when a positive voltage is applied to the p-type region and a negative voltage is applied to the n-type region. It is then said to be *forward biased*. If the polarities are reversed, the junction is reverse biased, and majority-carrier current cannot flow.

To be more precise, the junction allows current to flow when the forward bias imparts to the electrons the minimum energy they require to travel from the n side

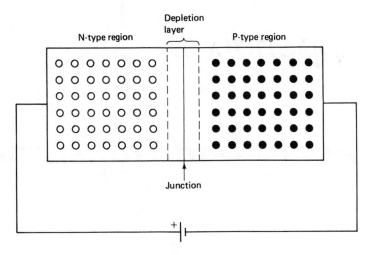

Figure A-1   Semiconductor diode.

to the *p* side of the depletion region. This is called *barrier energy* and is approximately 0.65 V for silicon and 0.25 V for germanium.

A *bipolar transistor* (BJT) is shown in Figure A-2. As you can see, it consists in effect of two diodes connected back to back. That is, they are arranged in an *np–pn* sequence. They also could be arranged in a *pn–np* sequence. However, instead of being separate diodes, they are combined in an *npn* or *pnp* configuration. The included *p* or *n* region is also made very thin. In Figure A-2, the *npn* transistor is shown connected to batteries so that the left "diode" is forward biased and the right "diode" is reverse biased. Majority carriers flow readily across the left junction, but not across the one on the right. In the *n* region on the left, electrons are majority

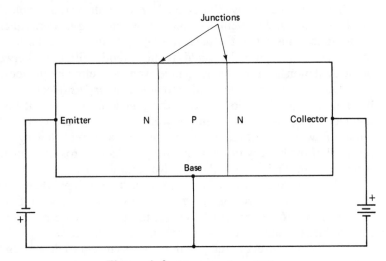

Figure A-2   Diagram of npn BJT.

carriers, so they are "injected" by the forward bias into the *p* region. However, in the *p* region electrons are minority carriers, and because this region is very thin, they soon find themselves in the depletion region of the other, reverse-biased junction. This results in the electrons being swept through this junction into the right *n* region. In a *pnp* transistor, the action is similar, but the majority carriers injected into the base are holes, so the biasing voltages have to be the other way round.

For reasons that are now obvious, the region that provides the majority carriers is called the *emitter*, and the region that receives most of them is called the *collector*. The middle region is called the *base*, because in early transistors it actually was (see Figure A-3).

Field-effect transistors (FETs) are unipolar devices, in which only one type of charge carrier is used. There are two types: the *junction field-effect transistor* (JFET), and the *metal oxide semiconductor field-effect transistor* (MOSFET).

Figure A-4 shows an *n*-channel JFET formed from a barrel-shaped slice of *n*-type silicon into which two *p*-type regions are diffused. Connections are made to the ends of the channel, which are called the *source* and the *drain*, and to the *p* regions, which are called the *gate*.

When a voltage is applied as shown, electrons flow through the channel from the source to the drain. If a negative voltage is applied to the gate, the junction between the gate and the *n*-channel will be reverse biased, and a depletion layer is formed. The gate has a higher proportion of impurities than the channel, so the depletion layer extends into the channel rather than into the gate. The higher the voltage on the gate is the farther the depletion layer extends into the channel. This is therefore called a *depletion device*.

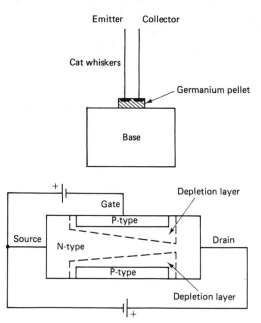

Figure A-3 Earliest transistors. A germanium pellet was mounted on a metal base; two thin metal leads called "cat whiskers" made contact with it. Small regions of the germanium around the points of the leads were of opposite polarity to that of the rest. This kind of transistor (now outmoded) was called a point-contact transistor.

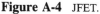

Figure A-4 JFET.

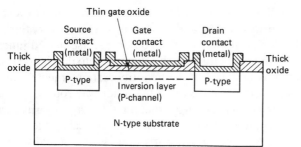

Figure A-5 MOSFET.

Since the depletion layer is virtually devoid of charge carriers, channel-current flow is decreased in proportion to the voltage on the gate. Thus the gate voltage controls the channel current. If the gate voltage is sufficiently negative, the depletion layer will extend across the whole of the channel and prevent any current flow. The value of gate voltage at which this occurs is called the *pinch-off voltage* (V_p).

A MOSFET differs from a JFET in not having a junction between the gate and the channel. As shown in Figure A-5, the most widely used type of MOSFET consists of successive layers of metal, oxide, and semiconductor material. The source and drain are formed by diffusing impurities into the substrate to make regions of opposite type. When a voltage of proper polarity is applied to the gate, its electric field penetrates the insulating layer and the substrate, repelling like charge carriers and creating an inversion layer, or channel, between the source and the drain regions that is of the same polarity as they are. Thus current can flow between them. The size of the current depends on the depth of the channel, which is proportional to the voltage on the gate. Therefore, a MOSFET with no voltage on its gate does not conduct, and as its conduction increases with the gate voltage, it is called an *enhancement device*. It takes a certain minimum voltage to form the channel, which is called the *threshold voltage* (V_T). The channel is designated *p* or *n* according to whether the channel current consists of holes or electrons. An *n*-channel device has positive drain and gate voltages; a *p*-channel device has negative drain and gate voltages.

It is possible also to make a depletion MOSFET, which works in a similar manner to a JFET. This type is normally conducting, and a sufficiently high gate voltage will pinch off the channel.

Appendix C

Converting a Breadboard Experiment to a Permanent Form

You may want to convert your breadboard power supply or function generator into a permanent piece of test equipment. This is easily done with a *perfboard*, or predrilled printed circuit board. These are printed circuit (PC) boards punched with holes spaced 0.1 inch apart, the same as on your breadboard. On one side each hole may have a ring of copper or solder to which a component lead may be soldered. PC boards are available in electronic parts stores.

You will only require sections of board about the same size as the area on the breadboard occupied by your experimental circuit. It is a simple matter to transfer each component from the breadboard to a corresponding position on the PC board, threading its leads through the appropriate holes. Keep all the components on the side of the board opposite the copper or solder rings, and then connect their leads, on the other side, to wire the circuit. Instead of mounting an IC directly on the board, it is safer to use an IC socket, which cannot be affected by the heat of soldering. The IC can be plugged into it afterward. Observe the soldering precautions given in Chapter 1.

You can give the function generator its own built-in power supply, mounted on the same PC board, and using the one given in Chapter 2, except that you will not need the -12-V_{dc} supply. Alternatively, you can provide terminals for an external power supply.

You should obtain a cabinet of the proper size for this board. Cabinets of various sizes for home-brew devices are also available in electronic parts stores. Some are sold together with PC boards, assembly hardware, protective feet, and even labels. It will serve as a housing for the PC board and as a means for mounting the controls and input and output terminals. Get a box just large enough to contain

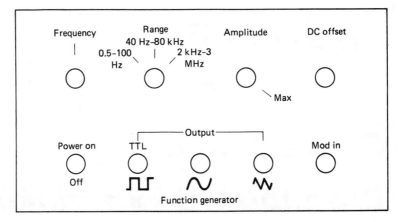

Figure A-6   Possible layout for the control panel of a function generator. Symmetry and waveform controls are internal (screwdriver) controls, which are adjusted during calibration only. (No controls are shown for FSK or modulation percent; these can be included if desired.)

the board and the terminals and panel controls (if any). Keep leads as short as possible to prevent pickup or radiation of unwanted signals.

The controls for frequency and range described in Chapter 3 should be provided by a potentiometer and a rotary switch mounted on the front panel of the box, with suitable knobs. The other controls are also potentiometers, plus a toggle switch for power if you include a power supply. The input and output terminals should be BNC connectors. A possible layout is shown in Figure A-6, but you can design your own, of course.

Glossary of Electronic Terms Used in This Book

AC   Alternating current.

Active filter   Electronic filter with one or more operational amplifiers (does not use inductors).

Alternating current   Electric current that reverses direction periodically, usually many times per second.

Ampere   Unit of electric current that is produced by an EMF of one volt in a resistance of one ohm.

Amplifier   Circuit with an active element that uses external power to boost the amplitude of an input signal, while preserving its essential characteristics.

Amplitude   Maximum level of an audio or other signal in terms of its current or voltage.

Analog   Term used to describe an electrical signal that varies continuously in amplitude or frequency in correspondence with some other physical variable (such as an audio signal that varies correspondingly with sound).

AND gate   Logic gate that gives a true output when all inputs are true.

Astable multivibrator   Free-running multivibrator (does not require a trigger); type of *RC* oscillator.

Bandpass   Range of frequencies passed with negligible attenuation by a circuit (also called passband).

Bandstop   Range of frequencies rejected by a network.

Base   Region between collector and emitter of transistor.

Battery   Two or more cells comprising a source of dc (also said of a single cell).

Bias   Direct voltage applied to a device, such as a transistor, to establish its level of operation.

Binary digit   The digits 1 and 0 of the binary number system.

Bipolar junction transistor   Transistor with two junctions.

Bit   Binary digit (1 or 0).

BJT   Bipolar junction transistor.

Block diagram   Diagram showing the sections or stages of an electronic system, connected by lines or arrows, but not the circuit components.

Bridge rectifier   Circuit in which four rectifier diodes are connected in a bridge circuit so that ac applied to the input becomes dc at the output.

BT   Class letters for battery.

Buffer   Term describing anything that prevents undesired interaction between two circuits or components (buffer amplifier, buffer capacitor, and so on).

Bypass capacitor   Capacitor to provide path for ac around a resistor to avoid undesired negative feedback.

C   Class letter for capacitor.

Capacitance   Ratio between the amount of electric charge that has been transferred from one plate of a capacitor to the other plate and the resulting voltage difference between them ($C = Q/V$), expressed in farads.

Capacitor   Device consisting essentially of two metal plates insulated from each other by dielectric material that can store an electric charge, block dc, and pass ac.

Cathode   1. Terminal of a diode that is more negative than the other terminal (anode) when the diode is forward biased. 2. Positive terminal of a battery

Cathode-ray tube   Funnel-shaped vacuum tube in which the wide end is coated with a phosphor that glows where struck by a beam of electrons projected from an electron gun in the narrow end, the beam being controlled by deflection plates or coils (a picture tube is a form of cathode-ray tube).

Choke   Inductance that presents considerable impedance to higher frequencies while offering little resistance to dc.

Clipper   Circuit that limits all signals to a fixed maximum value.

Collector   In a bipolar junction transistor, the electrode into which the majority carriers flow.

Color coding   Using colors to identify terminals, wires, and the like, or to show values of some components (for example, resistors).

Common emitter   Transistor amplifier circuit in which the emitter is common to both the input and output circuits (the most widely used type of transistor amplifier).

Common-mode rejection   Cancellation, in a differential amplifier, of signals that are common to both input terminals.

Comparator   Circuit that compares two signals and generates a difference, or error signal, if they are not identical.

Constant-current source   Circuit that resists changes in current, often by the use of a high-value resistor that does not allow much variation in the current flowing through it.

CR   Alternative class letters for semiconductor diode (D is preferred).

Crystal   Natural or synthetic piezoelectric or semiconductor material, whose atoms are arranged in a geometric lattice and which (in the case of quartz) vibrates at a fixed frequency according to how it is cut.

Current amplifier   Power amplifier.

D   Class letter for semiconductor diode.

dB   See *decibel*.

Decibel   One-tenth of a bel, logarithmic unit for expressing ratio between two amounts of power, voltage, or current [$dB = 10 \log_{10}(P_1/P_2)$. $dB = 20 \log_{10}(V_1/V_2)$. $dB = 20 \log_{10}(I_1/I_2)$].

Degeneration   Negative feedback.

Demodulation   Recovery of modulating signal from carrier.

Dielectric   Insulating medium between two plates of a capacitor.

Differential amplifier   Amplifier having two similar input circuits, connected so that they respond to the difference between two applied signals, but suppress signals that are common to both.

Digital circuit   Circuit that operates like a switch (it is either on or off).

DIP   See *dual-in-line package*.

Drain   Region of an FET that receives electrons or holes.

Dual-in-line package Integrated circuit packaged in a plastic case about $\frac{3}{4}$ inch long and $\frac{1}{3}$ inch wide, with pins along each long side, for mounting on a printed circuit board.

Electron Subatomic particle carrying a single basic charge of electricity; present in the outer shells of all atoms, and also free.

Emitter In a bipolar junction transistor the region that injects charge carriers into the base when the emitter–base junction is forward biased.

Enhancement mode Mode of operation of MOS field-effect transistor that is normally off with zero gate voltage.

Exclusive NOR logic gate Inverted exclusive OR logic gate.

Exclusive OR logic gate OR logic gate in which the output is false when both inputs are true.

Farad Unit of capacitance (usually with prefixes micro- or pico-).

Feedback In an amplifier, return of a fraction of the output to the input; may be negative (reduces gain, promotes stability) or positive (increases gain, promotes oscillation).

FET Field-effect transistor.

Field-effect transistor Transistor in which either electrons or holes (depending on the semiconductor material used) flow from the source to the drain via a channel controlled by the gate (see insulated-gate field-effect transistor and metal oxide semiconductor field-effect transistor).

Filter Selective network that passes some frequencies while blocking or attenuating others.

Flip-flop Bistable multivibrator with two states that switches from one to the other upon application of a trigger signal.

FM See *frequency modulation*.

Free-running multivibrator Astable multivibrator.

Frequency modulation Modulation of a sine-wave carrier so that its frequency at any given moment differs from the carrier frequency in proportion to the amplitude of the modulating frequency.

Full-wave rectifier Rectifier circuit using both positive and negative excursions of the power-line ac to produce dc.

Gain In an amplifier, the amount by which the output signal exceeds the input signal.

Gate 1. Logic circuit in which the output signal is determined by the signals on two or more inputs. 2. Control electrode in a field-effect transistor.

Ground In a circuit, a point with zero potential, often connected to earth, that serves as the voltage reference.

Half-wave rectifier Rectifier circuit that uses only the positive (or negative) excursions of the power-line ac to produce dc.

Hertz Unit of frequency (one cycle per second).

Hole Gap in the valence shell of a semiconductor atom caused by the removal of an electron, generally treated as if it were a "positive electron."

I Symbol for electric current.

IC See *integrated circuit*.

IGFET Insulated-gate field-effect transistor.

Insulated-gate field-effect transistor See *metal oxide semiconductor field-effect transistor*.

Integrated circuit Small piece of semiconductor (usually silicon) upon which transistors and other circuit elements have been fabricated by deposition or diffusion and connected together by deposited metal conductors to form complex circuits or systems that offer great advantages in terms of small size, economy, and reliability.

Inverter Common-emitter amplifier used to reverse the polarity of a signal (sometimes called a NOT gate).

JFET See *junction field-effect transistor*.

Junction Boundary between p and n regions in a transistor or diode.

Junction field-effect transistor Field-effect transistor in which the gate is insulated from the channel by a reverse-biased junction.

LCD See *liquid-crystal display*.

LED See *light-emitting diode*.

Light-emitting diode Solid-state diode that emits light when forward biased.

Linear Having an output that varies in direct proportion to the input.

Liquid-crystal display Digital display consisting of two sheets of glass separated by sealed in, normally transparent, liquid-crystal material; the outer surface of each glass sheet has a transparent conductive coating such as tin oxide or indium oxide, with the viewing side coating etched into character-forming segments that have leads going to the edges of the display; a voltage applied between the front and back electrode coating disrupts the orderly arrangement of the molecules, darkening the liquid crystal to form visible characters.

Logic circuit Circuit that responds to two or more digital inputs to give a digital output that is a logical function of the inputs (also called a gate).

Memory   Device in a computer that stores data. Internal memory (random-access memory) is temporary storage; external memory (consisting of diskettes, tape, and so on) is permanent storage; internal read-only memory is permanent storage.

Metal oxide semiconductor field-effect transistor   Field-effect transistor in which the gate is insulated from the channel by a film of silicon dioxide.

Meter   Any measuring device, but usually meaning one with a dial and a pointer (for example, a voltmeter, ammeter, or ohmmeter).

Microprocessor   Control and processing portion of a computer, usually an integrated circuit.

Modulation   Modulating some characteristic of a carrier wave so that it carries video or audio information, usually as a TV or radio wave.

MOSFET   Metal oxide semiconductor field-effect transistor.

Multiplexer   Device for combining two or more signals for transmission via a single channel, without losing their separate identities.

Multivibrator   *RC* oscillator using two transistors that conduct alternately.

NAND gate   Inverted AND logic gate that gives a true output when all inputs are false.

Negative feedback   Feedback where part of the output signal is fed back 180 degrees out of phase to the input (also called degeneration or inverse feedback).

NOR gate   Inverted OR logic gate that gives a true output when any input is false.

Ohm   Unit of resistance (potential of one volt causes a current of one ampere to flow through a resistance of one ohm).

Operational amplifier   Stable high-gain dc amplifier that depends on external negative feedback to determine its functional characteristics (usually in the form of an integrated circuit).

OR gate   Logic gate that gives a true output when any input is true.

Oscillator   Amplifier employing positive feedback to generate an ac signal whose frequency is determined by the circuit components.

Phase angle   Angular relationship between current and voltage in an ac circuit.

Phase detector   Circuit that compares the phase of an input signal with that of an oscillator and, if they are different, generates an error signal to adjust the oscillator frequency to agree with that of the input signal.

Phase inverter   Amplifier that changes the phase of a signal by 180 degrees.

Phase-locked loop Circuit similar to a phase detector; it contains a voltage-controlled oscillator whose frequency is controlled by the phase detector's error signal (this circuit has many uses, one of which is FM detection).

Pi Greek letter π, which designates the ratio of the circumference of a circle to its diameter (=3.1416 approximately).

PLL Phase-locked loop.

Polarized capacitor Electrolytic capacitor in which the dielectric film is formed next to only one plate, so it is likely to break down if an oppositely polarized voltage is applied.

Positive feedback Feedback where part of the output is fed back in phase with the input.

Potentiometer Variable resistor with continuously adjustable sliding contact, used as a voltage divider (commonly called a "pot").

Power supply Source of dc to operate active elements (may be a battery or circuit to convert line voltage to dc).

Preferred values Widely used series of values for resistors and other components to reduce the number of different sizes that must be kept in stock.

Q Class letter for transistor.

Quartz crystal See *crystal*.

R Class letter for resistor.

Reactance Opposition to ac offered by capacitance or inductance.

Rectifier Device, such as a diode, that converts ac into unidirectional current.

Regeneration Positive feedback.

Register Set of flip-flops for temporary storage of data.

Regulation In a power supply, the ability to maintain a constant output regardless of variations in line voltage or load impedance.

Resistance Property of a conductor that determines the current that flows in it as a result of the application of a given voltage.

Resistor Device that introduces a required resistance into a circuit.

Rheostat Variable resistor with continuously adjustable sliding contact for controlling current.

RTL Resistor–transistor type of logic gate.

S Class letter for switch.

Saturation In a transistor, when an increase of base current produces no further increase of collector current.

Sawtooth wave Triangular-shaped waveform, in which the rising portion is slower than the falling portion, resembling the teeth of a saw.

Semiconductor Material such as silicon or germanium with a conductivity between metals and insulators, used for transistors, diodes, and the like.

Series Connecting of components end to end in a circuit to provide a single path for the current.

Shunt Connecting of any component in parallel with another.

Signal Visible, audible, or other conveyor of information.

Source Region of an FET that emits electrons or holes.

Substrate Physical base on which a microcircuit is fabricated, especially a silicon chip.

Switch Mechanical or electrical device that completes or breaks the path of a current.

T Class letter for transformer.

Thermistor Type of resistor whose resistance decreases as its temperature increases.

Time constant In a capacitor–resistor circuit, the time in seconds required for the capacitor to reach 63.2% of its full charge after a voltage is applied.

Transformer Device consisting of inductively coupled coils that transfers ac between circuits, in some cases changing the voltage and current values.

Transistor Active semiconductor component used as an amplifier, switch, or the like.

TTL Transistor–transistor logic.

U Class letter for integrated circuit.

V_{CC} Symbol used in schematic diagrams to designate the collector power supply in a bipolar transistor circuit.

VCO Voltage-controlled oscillator.

V_{DD} Symbol used in schematic diagrams to designate the drain power supply in a field-effect transistor circuit.

Voltage-controlled oscillator *RC* oscillator in which, for instance, the time of charging a capacitor through a resistance (and thereby changing the oscillator frequency) may be varied by changing the bias voltage applied to a transistor in series with the resistance.

Waveform   Shape of an electromagnetic wave (graph of its amplitude plotted against time).

Wien bridge oscillator   *RC* oscillator employing an operational amplifier.

XNOR   Exclusive NOR logic gate.

XOR   Exclusive OR logic gate.

Y   Class letter for crystal.

Zener diode   Diode having the characteristic that it becomes conductive (nondestructive breakdown) at a certain reverse-bias voltage, so it can be used as a reference voltage.

Index

Index

A

ac. *See* Alternating current
Active filter, 122–133
ADC. *See* Analog-to-digital converter
Alternating current (ac), 12
Analog circuit, 73, 74, 115
Analog-to-digital converter (ADC), 134–140
  dual-slope converter, 136
  simultaneous converter, 134, 135
  staircase-ramp converter, 136
  successive-approximations converter, 137–140
AND gate, 77
Antilog amplifier, 124

B

Binary number code, 45–47
Bipolar junction transistor (BJT), 150–157

BJT. *See* Bipolar junction transistor
Breadboard, solderless, 1–3, 29, 38
Bus, 106

C

Capacitance, 6
Capacitor, 6, 7
Capacitor test, 31, 32
Cathode-ray tube (CRT), 54
Clock, 60ff, 66–69
CMOS. *See* Complementary metal oxide semiconductor logic
CMR. *See* Common-mode rejection
Color code, 3–7
Common-mode rejection (CMR), 116
Comparator, 65
Complimentary metal oxide semiconductor logic, 80–82
Counter, 93
  asynchronous binary ripple counter, 93–95
  decade counter, 97, 98

Counter (*cont'd*)
  integrated counters, 98
  synchronous counter, 95
CRT. *See* Cathode-ray tube

D

DAC. *See* Digital-to-analog converter
dc. *See* Direct current
Decoder, 49, 50
Demultiplexer, 113, 114
Difference amplifier, 122
Differential amplifier, 115, 116
Digital circuit, 73ff
Digital-to-analog converter (DAC),
  140–146
Diode, 7, 149, 150
DIP. *See* Dual-in-line package
Direct current (dc), 12
Dual-in-line package (DIP), 3
Duty cycle, 40

E

ECL. *See* Emitter-coupled logic
EEPROM. *See* Memory
Emitter-coupled logic (ECL), 79
Encoder, 47–49
EPROM. *See* Memory

F

FET. *See* Field-effect transistor
Field-effect transistor, 151, 152
Filter, 16
Filter capacitor, 16–18, 20
Flip-flop, 89, 91–93
Full-wave rectifier, 18–20
Function generator, 33–43

G

Gate, 76, 77, 82

H

Half-wave rectifier, 15, 16

I

IIL, I^2L. *See* Integrated injection
  logic
Integrated circuit, 3, 8–10, 34ff
Integrated injection logic, 79, 80
Integrator, 33, 34
Inverter, 76

J

JFET. *See* Field-effect transistor
Junction, 149, 150

L

Latch, 89, 90
LCD. *See* Liquid-crystal display
LED. *See* Light-emitting diode
Light-emitting diode (LED), 44–51
Liquid-crystal display (LCD), 51–53
Logarithmic amplifier, 123
Logic clip, 85
Logic families, 77–82
Logic probe, 82–84
Logic pulser, 85

M

Memory, 102–114
  random-access memory (RAM),
    104–106

register, 102–104
read-only memory (ROM), 107
semiconductor types, 107
timing chart, 110
Modulation, 41
MOSFET. *See* Field-effect transistor
Multiple-character display, 54, 55
Multiplexer, 111, 112
Multiplier, 33, 35, 124, 125

N

NAND gate, 60–62, 77–79
NOR gate, 77
NOT gate. *See* Inverter
n-type semiconductor, 148

O

Operational amplifier (op-amp),
    116–133
  active filter, 128–133
  antilog amplifier, 124
  buffer, 120
  comparator, 120, 121
  difference amplifier, 122
  frequency compensation, 121
  inverting voltage amplifier,
      118–119
  logarithmic amplifier, 123
  noninverting voltage amplifier,
      119–120
  offset compensation, 121
  pulse generator, 126, 127
  sinusoidal oscillator, 126
  square-wave generator, 125, 126
  summing amplifier, 122, 123
  voltage multiplier and divider, 124,
      125
Optical isolator, 55, 56
Optoelectronic devices, 44–59
OR gate, 77

P

PCB. *See* Printed-circuit board
Peak-inverse voltage (PIV), 15
Perfboard, 153
PIV. *See* Peak-inverse voltage
Potentiometer (pot). *See* Resistor,
    variable
Power supply, 12–32
Printed-circuit board (PCB), 153
PROM. *See* Memory
p-type semiconductor, 148
Pulse generator, 40, 127

R

RAM. *See* Memory
Rectification, 14–20
Rectifier, 14, 15, 18–20
Register, 102–104
Regulation, 20–21
Regulator, 21–28
Resistive-ladder network, 141
Resistor
  fixed, 3–5
  variable, 6
Rheostat, 6
Ripple. *See* Filter capacitor
rms. *See* Root-mean-square voltage
    or current
ROM. *See* Memory
Root-mean-square (rms) voltage or
    current, 3

S

Semiconductor, 147, 148
Semiconductor devices, handling,
    10, 11
Seven-segment numeric display,
    49–51
Square-wave generator, 33, 125, 126

Summing amplifier, 122, 123
Surge resistor, 18
Switching regulator, 27, 28

T

Three-terminal regulator, 25–28
Timer, 63
Timing-pulse generator. *See* Clock
Transformer, 13, 14
Transistor, 8, 150–152
Transistor-transistor logic (TTL),
     78–79
Troubleshooting
   A/D and D/A converters, 146
   clocks, 69–72
   counters, 99–101
   digital logic, 85–88
   function generator, 42–43
   operational amplifier circuits,
        57–59
   power suppliers, 28–32
   registers, 114

Truth table, 75, 76
TTL. *See* Transistor-transistor logic

V

Vacuum-fluorescent display, 53, 54
Valence electrons, 147
VCO. *See* Voltage-controlled
     oscillator
Voltage comparator. *See* Comparator
Voltage-controlled oscillator, 35, 69

W

Wien-bridge oscillator, 126

Z

Zener diode, 21
Zener regulator, 21–28